THE LEAK

THE LEAK

**Politics, Activists, and Loss of Trust
at Brookhaven National Laboratory**

ROBERT P. CREASE

with Peter D. Bond

**The MIT Press
Cambridge, Massachusetts
London, England**

The MIT Press would like to thank the anonymous peer reviewers who provided comments on drafts of this book. The generous work of academic experts is essential for establishing the authority and quality of our publications. We acknowledge with gratitude the contributions of these otherwise uncredited readers.

This book was set in Adobe Garamond and Berthold Akzidenz Grotesk by Westchester Publishing Services. Printed and bound in the United States of America.

Library of Congress Cataloging-in-Publication Data

Names: Crease, Robert P., author.
Title: The leak : politics, activists, and the loss of trust in science at
    Brookhaven National Laboratory / Robert P. Crease with Peter Bond.
Description: Cambridge, Massachusetts : The MIT Press, 2022. | Includes
    bibliographical references and index.
Identifiers: LCCN 2021057607 | ISBN 9780262047180 (hardcover)
Subjects: LCSH: Brookhaven National Laboratory—History. | Science and state—
    United States. | Radioactive decontamination—New York (State)—Long Island.
Classification: LCC QC789.U62 B73 2022 | DDC 363.17/990974725—
    dc23/eng20220630
LC record available at https://lccn.loc.gov/2021057607

10 9 8 7 6 5 4 3 2 1

Contents

PROLOGUE

In 1997, a leak of water containing radioactivity was discovered at Brookhaven National Laboratory. Though federal, state, and local officials declared that it posed no health hazard either to the lab's employees or to the surrounding community, it triggered a media and political firestorm. In its wake came a startling series of events that included the firing of Brookhaven's corporate manager, changes in the lab's leadership, transformation of the lab's culture, significant alterations to management contracts of Department of Energy scientific facilities, the permanent closure of the lab's research reactor, and even calls to close the lab itself. While 1997 was the lab's fiftieth anniversary, it was a year of chaos rather than celebration.

A quarter of a century later, in a period of social and political uncertainty when the value and authority of science are urgently needed, it is vital to look back on this episode. Amid the unruly happenings one can make out fault lines between scientists, politicians, media, and the public that have only grown larger and more intractable. The story of the events at Brookhaven in 1997 exposes dangers that US science faces now and in the foreseeable future and illustrates lessons about the management of large scientific facilities that need to be learned or relearned. This book is therefore about not just Brookhaven National Laboratory but the kinds of social and political dynamics on which the existence of scientific facilities such as Brookhaven continue to depend.

Though the health and safety impact of Brookhaven's leak was negligible, activists and the media at times applied sensationalist language to the episode. While such language would have been appropriate to Chernobyl, Bhopal, and the Challenger launch disasters, it was not to the BNL leak. The story therefore helps shed light on current controversies where suspicions and fears, combined with loss of trust and political ambitions, strongly shape controversies involving issues with scientific and technical dimensions, such as the safety of vaccines, the reality of climate change, and the dangers of toxic chemicals, for starters.

We were there. Crease, a Stony Brook University faculty member, is a philosopher and historian of science who was then finishing a book on the lab's first half-century.[1] Bond had worked at the lab for twenty-five years and began 1997 as chair of the Physics Department; he was then proposed as the lab's interim director twice during the year and appointed once; he has written an article about his personal experiences.[2] (Both are referred to here in the third person.) This makes us interested parties who might be viewed as biased in presenting the story. Then and now, we found many of the unfolding events to be like the plot twists in some outrageous tragicomedy whose spectrum of vivid characters ranged from cultivated to clownish. However, the passage of more than twenty-five years, together with our subsequent experiences, have allowed us, we hope, to look back more insightfully and with fresh eyes. We also had unprecedented access to many participants, internal emails, and the minutes of key meetings, which put us in a special position to observe and judge the actors, events, and dynamics.

Despite our roles, we have striven to make our account, based on data, documents, and interviews, not simply one side or perspective to be set among and against others but a picture in which, we hope, all participants can recognize themselves—and we hope it demonstrates the lessons that everyone concerned about the present-day threats to human health and welfare need to draw from the episode.

1 THE LEAK DISCOVERED

Radioactive pollution does not have to be injurious to health to be socially undesirable.
—BERNARD MANOWITZ, Brookhaven National Laboratory, 1949

TRITIUM

Late on January 8, 1997, Doug Paquette, a hydrogeologist at Brookhaven National Laboratory (BNL), received an analysis of groundwater samples near the lab's High Flux Beam Reactor (HFBR). The analysis showed a level of tritium, a radioactive form of hydrogen, above the Environmental Protection Agency's drinking water standard.[1] He shot back a request to confirm the results, adding that "this is/could be a very tricky situation."[2]

Brookhaven National Laboratory is a world-famous scientific institution located in New York's Suffolk County in eastern Long Island. It had been established in 1947 by a contract between a consortium of nine prominent universities in the Northeast—known as Associated Universities, Inc. (AUI)—and the Atomic Energy Commission (AEC) as a place to bring the resources of American academia and government together to build and operate scientific instruments whose scale was too big for individual universities to afford. The lab's major instruments would include particle accelerators and nuclear reactors. The first reactors had been built for military purposes as a part of the Manhattan Project, but reactors were also essential for nonmilitary scientific research throughout virtually all areas of science. The Department of Energy (DOE) eventually became the owner of the US national labs, so that in 1997 Brookhaven was owned by the DOE and operated by AUI, subject to the DOE's regulations and oversight.[3]

Brookhaven's research spanned many fields, including physics, chemistry, biology, engineering, medicine, and others. By the beginning of its half-centennial in 1997, the lab's research had resulted in four Nobel Prizes (now seven), more than any other laboratory worldwide in a comparable period.

Yet despite Brookhaven's eminence, it had become vulnerable. In the wake of Paquette's finding, an array of elements—Suffolk County's history, Long Island's geology, environmental concerns, and changing demographics, combined with poorly considered actions by lab and government officials, an antinuclear movement, political exposure, media coverage, celebrities, and the challenge of risk communication even in the best of circumstances—led to the dismissal of AUI, the closure of the HFBR, and calls to close the lab itself.

* * *

For a few days, neither the public nor local politicians knew of Paquette's findings as he rechecked them. When he got confirmation, he alerted his bosses in the Environmental Health and Safety (ES&H) division, and scheduled an emergency meeting to draw up an "Occurrence Report" as required by the DOE.[4] His bosses scheduled a joint meeting with the lab's Reactor Division for Monday morning. Before leaving work on Friday, Paquette used his knowledge of the groundwater flow to devise a plan to install more monitoring wells both to find the tritium source and to see if it was an active leak or previous spill. It would be the last weekend for months that many employees in the lab's administration, ES&H, and media relations would spend at home.[5]

On Monday morning, January 13, the ES&H group reviewed the discovery. It did not appear to pose a danger to health from a scientific perspective. It was deep underground, and tritium, a form of hydrogen whose nucleus consists of a proton and two neutrons, emits a weak form of radiation (beta decay) that does not penetrate skin and travels only an inch in air; a sheet of paper stops its radiation. Tritium combines with oxygen to make water, and is used in a variety of ways: biomedical researchers use it as a tracer, hydrologists put it in groundwater to trace the flow, and oceanographers inject it into the sea to track currents. It is widely used in self-illuminating "Exit" signs and luminous watches. Tritium in the groundwater is a health hazard only if humans drink it, and then only at a high enough concentration in sufficient quantities. The drinking water standard set by the EPA[6] is based on a limit equivalent of a person drinking two liters of water at 20,000 picocuries per liter per day per year (14,600,000 pCi total per year). In this case, the scientists were alarmed because tritium was found in monitoring wells on-site near the reactor, and, more importantly, because its origin was unknown. The group brainstormed for ideas about the possible source, and assigned individuals to investigate each.[7]

Some members of the group thought it too early to spread the news. Bill Gunther, head of the BNL Office of Environmental Restoration (OER), was one. The data were sparse, the tritium was no health hazard, and as a Superfund site since 1989 the lab's environmental restoration staff had far more pressing problems to worry about involving certain chemical contamination issues—though these, too, did not pose a danger to health. Tritium's half-life is 12.3 years, and it would spread and dilute in the slow-moving groundwater, all but vanishing by the time it reached the lab's border a mile and a half away.[8]

Nevertheless, the group alerted lab director Nicholas Samios, who asked the group a series of questions: Do you believe the numbers? (Yes.) What's the source? (We don't know.) Are you sure it's not the reactor vessel? (Yes.) Are you sure it's the reactor? (No.) Are you sure it's not the reactor drains? (No.) What are other possibilities? (The spent fuel pool, some source upstream of the reactor.)

The HFBR happened to be shut down for routine maintenance, and Samios decided to keep it shut until the source was found and stopped. He ordered Paquette to implement his plan to install more wells, and notified Martha Krebs, head of the DOE's Office of Energy Research (now called the Office of Science) in Washington, D.C. Brookhaven was seeking to upgrade the HFBR, and among other things Samios wanted to be sure the DOE appreciated the lab's concern for safety and the environment.[9] Krebs ordered Samios to get DOE approval for any press releases on the subject. Such

Figure 1.1
Brookhaven National Laboratory circa 1997 showing the Relativistic Heavy Ion Collider (RHIC) ring at top, the domed High Flux Beam Reactor (HFBR) to the right of the stack, the Brookhaven Graphite Research Reactor (BGRR), to the left of the stack, and the National Synchrotron Light Source (NSLS), the white building below the stack with a circular extension.

approvals would each take days, for they would have to go up the bureaucratic chain from the DOE's on-site office, through its Chicago Operations Office that oversaw numerous labs including BNL, and only then to the DOE's Washington, D.C., office and to its secretary, a Cabinet official.[10]

* * *

Two days later Paquette's first new monitoring wells—called Geoprobes after the company that made them—were pounded into the HFBR lawn. Data from these would allow him to plan a second and third round to pin down the tritium's location and extent, in what would soon comprise a network of 24 monitoring wells.[11]

Meanwhile, lab officials distributed data, maps of wells and water contours, and a list of intended actions to federal, state, and local regulatory agencies.[12] Lab staff combed the HFBR's records; began testing pipes, drains, sewers, and the pool where the reactor's used fuel rods were stored; and put together a plan for how to plot the extent of the tritium, deal with it, and distribute the information.[13]

On Thursday, Mona Rowe, the lab's head of media relations, left for a prearranged luncheon with the Southampton Rotary Club, which had asked her to talk about the lab's legacy environmental issues. As the DOE had still not approved release of the news, she could say nothing about the tritium discovery.

The next day, Samios sent a memo to employees about the finding, assuring them that it didn't affect drinking water or safety. That Friday happened to be a bittersweet moment for the lab's media and communications staff, as that evening Janet Sillas, its flamboyant office administrator of twenty-three years and a talented singer, was performing at her own retirement party. Rowe, another talented musician, was accompanying her on piano. In the middle of Sillas's signature song, "Broadway Baby," a woman from the office of Sue Davis, Associate Director for Reactor Safety & Security, slipped into the room, placed a document on the piano music rack in front of Rowe, and whispered, "The DOE-approved press release."

When the song ended, Rowe dashed to a phone to call Davis. Reporters know that Friday nights are the traditional time for organizations to announce bad news, and even with a preprogrammed media fax list the release would not get out until after 9:00 p.m. Fearing the optics, she convinced the DOE to let her postpone release until the next morning, which would give the media a full day to develop the story and contact sources. Rowe and her boss arrived at 7:00 a.m. on January 18, and handled press queries for the entire day.[14] A week had passed since Paquette had confirmed the tritium finding, and while all of the appropriate regulatory reporting had occurred, the public had only just been made aware. However, two congressmen—Senator Alfonse

D'Amato and Representative Michael Forbes—had not been alerted. This primed Paquette's "tricky situation" to erupt.

The next week, the ES&H staff frantically analyzed the Geoprobe data. Tritium was only found south in the groundwater from the reactor, so its source was evidently in that building. But it was difficult to form a clear picture with data coming in from different wells at different depths. At 7:00 p.m. on January 25, a baffled Bob Miltenberger wrote Paquette, "For those of you who thought more data would help us understand this issue, read these data tables and please help me understand what is going on." Paquette puzzled over the data and an hour before midnight wrote back, "The data does make some sense!!!!"[15] He outlined his view of the tritium's location, but cautiously warned that he needed more data from the probes that he had ordered installed that Monday morning, January 27.

Two days later, Davis, Gunther and others drew up and handed DOE officials their completed "Tritium Plume Recovery Plan," a comprehensive

Figure 1.2
Hydrogeologist Doug Paquette (standing) and two coworkers at a Geoprobe, with the HFBR in the background.

program to identify the size and source of the area with tritium, implement short- and long-term corrective actions, plan the eventual restart of the HFBR, and convey the information to the relevant people and agencies.[16] Davis began sending the local DOE office, every few days, extensive tables of the results of analysis of every well sample.[17]

On January 31, Samios sent lab employees a memo about the first phase of Paquette's test wells. Levels of tritium in the groundwater were eleven times the drinking water standard 100 feet south of the reactor building, dropping off farther away. More wells would soon give a better picture. Samios wrote that the EPA was independently analyzing the samples, and the data showed no contamination of drinking water or threat to either lab employees or neighbors.[18] Samios also set in motion a second phase of Paquette's plan to find the source. Samios, and most of the rest of the lab scientists, saw the issues posed by the tritium discovery as a simple matter: Look at the numbers, see what the harm is, and, if there's any, fix it.

But that's not how others saw the issues.

A HISTORY OF TROUBLED WATERS

"While nothing could seem simpler than a clear glass of water," writes James Salzman, "it is difficult to find a more complex or fascinating topic." Drinking water is infused with symbolism and myth, says Salzman, a professor of law and environmental policy at Duke University and author of *Drinking Water: A History*. Stigma can be attached to groundwater where the amounts of chemicals in it are so low that it doesn't matter, which makes whether it is safe or not difficult to discuss. "We need to find a way to discuss more honestly and openly what we mean by safe drinking water."[19]

The difficulty of discussing the safety of the groundwater in 1997 at Brookhaven National Laboratory was amplified by the presence of its unique geography, history, and personalities.

The geographical issue was that the island's drinking water comes from its uppermost aquifer, or water-containing mass of sand and gravel, deposited by a glacier, that sits over a layer of clay but below ground level.[20]

Containing around 70 trillion tons of water, it is Long Island's "sole source" aquifer, replenished by rain seeping in through the soil.

The historical issue was that, when Brookhaven opened in 1947, eastern Long Island was sparsely populated and inhabitants paid scant attention to groundwater purity. Ringed with beaches and fishing villages on the coast, the island's interior had a monotonous landscape of pitch pine and scrub oak, and various types of farms. The lab was built on the site of Camp Upton, a decommissioned Army camp in the middle of the island, used in both world wars; the remoteness of the location had been a reason for the location of both camp and lab.

But Suffolk County grew rapidly. Highways and railroads brought it within commuting distance of New York City and it developed and suburbanized, surpassing its neighbor Nassau County—about forty miles west of Brookhaven—as the fastest-growing suburban county in the country. Vacation homes for the wealthy sprang up, especially at the far end of Long Island's South Fork in towns collectively known as The Hamptons, about forty miles east of the lab. Right after World War II the county's population of about 250,000 mainly lined the shores, but half a century later the population of around 1,400,000 was scattered throughout the region.

With Long Island's development and industrialization came environmental issues—plumes of fertilizer, gasoline, pesticides, degreasers, and oil that slowly absorbed into the soil. Homeowners routinely disposed of solvents—now collectively known as volatile organic compounds (VOCs)—in the ground, and until the 1970s were encouraged to put "safe" degreasers and other solvents into cesspools. Service stations, factories, industries, and BNL used solvents to clean equipment, and they, too, frequently disposed of them by pouring them into the ground. At Camp Upton, solvents had been used to clean uniforms, tents, rifles, and other gear, and were mixed with paint. Other contaminants were dumped into the ground by Precision Concepts, an electronics fabrication company in an industrial park at the lab's southern boundary. Liquid contaminants would make their way into the groundwater, which moves slowly south about a foot per day at most; other contaminants would generally combine with soil and not move far. Fortunately, an aquifer is not an underground lake, where a drop of liquid immediately spreads, but more like a giant sandbox, where a drop flows down-gradient slowly in a plume and can in principle be cleaned up.[21]

In the 1970s, in an awakening of environmental sensitivity, the United States began to regulate solvents, degreasers, pesticides, and other chemicals. Brookhaven scientists shared this environmental concern: lab researchers were a driving force in founding the Environmental Defense Fund, and the EDF's certificate of incorporation (1967) was signed in a lab conference room. The lab scientists' environmental activism sometimes caused friction in neighboring communities.[22]

Many areas of Long Island, including the lab, had plumes that now exceeded acceptable levels. In 1989, the lab, joining a list of 20 other areas on Long Island, became a Superfund site, a program administered by the US Environmental Protection Agency (EPA) to identify and clean up areas of hazardous waste.[23] Part of the Superfund process involves surveys to identify and remediate legacy events and practices that resulted in environmental pollution, and the lab began a "Historical Legacy" survey to systematically track and clean pockets of VOCs and other contamination.[24] Some chemical plumes, including those from the Camp Upton days, were too deep— 180–200 feet or so down—to be a health hazard, for most people's wells reached down 60–80 feet. Tritium—a byproduct of normal operation of the HFBR—was regularly released in the lab's waste discharge into the Peconic River, which effectively originates on site, but at levels a fraction of the drinking water standard.[25] The lab began work on a state-of-the-art waste management facility and an upgrade of its sewage treatment plant to replace the ones in use since the lab was founded in 1947.

The lab also undertook a joint program with the Suffolk County Department of Health Services (SCDHS) to find chemicals that had gone off-site. At Brookhaven, each contaminated area at the lab was made a separate cleanup project, called an "operable unit," and six were ultimately created.[26]

A Superfund project involves a formalized set of procedures for investigation, cleanup, and community engagement that operated autonomously with respect to the other lab activities focused on science. Following those procedures, the lab established a community relations and outreach program, whose head was Kathy Geiger, a former journalistic researcher. Geiger

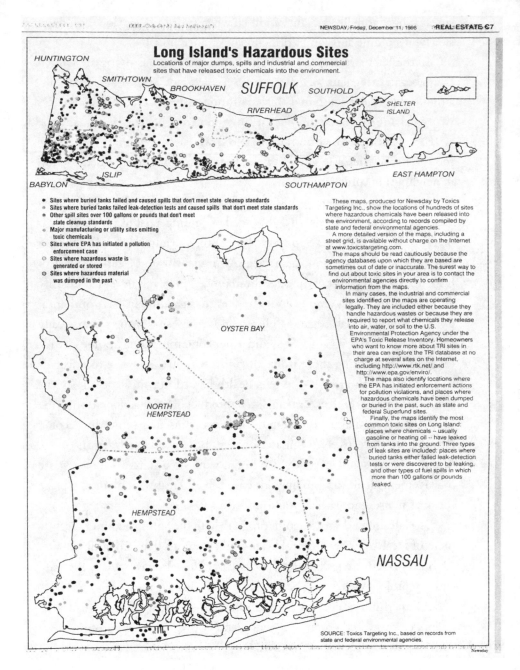

Figure 1.3

"Long Island's Hazardous Sites," *Newsday*, December 11, 1998. From Newsday. © 1998 Newsday. All rights reserved. Used under license.

reported to OER head Bill Gunther, who in turn reported to Associate Director for Reactor Safety & Security Sue Davis. The program issued quarterly public newsletters, set regular open meetings on cleanup practices, and established relationships with civic associations.

Besides the lab's Superfund status, three other developments had also ratcheted up environmental sensitivities on Long Island. One was an antinuclear movement sparked by an attempt to build a nuclear power plant on the North Shore. The second was reports that incidences of breast cancer on parts of Long Island were unusually high. The third was publicity surrounding cases, in other parts of the country, where industries had created and then covered up extensive chemical contamination causing incidences of cancer and other diseases.

Long Island's antinuclear movement was crystallized by a nuclear power plant that the Long Island Lighting Company began building in the late 1960s in Shoreham, a Suffolk County town ten miles north of the lab.[27] The recently passed National Environmental Policy Act (NEPA) mandated public hearings for such projects. Starting in 1976, local and national antinuclear activists staged highly effective protests at the hearings, targeted at newspaper and television audiences. Long Island environmental lawyer Irving Like, a key anti-Shoreham activist, advised that the hearings should become "a multi-media confrontation," saying, "Each day must become a dramatic and suspenseful event. . . . Phrase statements in a lively manner— which will read or sound well when recorded in the next day's press or radio account."[28] The protests succeeded in terminating the facility in 1989 as it was on the verge of operation. In the wake of the 1979 accident at Three Mile Island and the 1986 Chernobyl accident in the USSR—involving power reactors of 800 and 1,000 megawatts (MW), respectively—rumors of nuclear cover-ups and conspiracies acquired new currency and vibrancy. Brookhaven National Laboratory had two operating research reactors—the HFBR (30 MW) and the much smaller Brookhaven Medical Research Reactor (BMRR, 3 MW)—as well as the remnants of the closed and mostly decommissioned Brookhaven Graphite Research Reactor (BGRR, 32 MW). Reactors and radiation are associated with nuclear weaponry and accidents and arouse particularly strong fears and emotions. (When Rowe's sister asked how she could choose to work near a nuclear reactor, Rowe replied by asking how her sister could choose to live near the San Andreas Fault.[29]) These fears and emotions make it exceptionally challenging to establish and defend acceptable safety levels, especially when those establishing and defending them are from the institutions responsible for the radiation in the first place.

The second development occurred in the wake of a Long Island Breast Cancer Study initiated by New York State health scientists, which found that, from 1978 to 1982, the incidence of breast cancer in Nassau County was 12 percent above the New York State rate and 7.5 percent above the US rate, although in Suffolk County it was barely above average. This study fostered breast cancer activism, inspiring citizens to organize to collect information and compile maps of possible cancer clusters, and to pressure for more scientific studies about possible connections between the local environment and breast cancer. Rumors surfaced of cancer clusters, including one of a rare childhood cancer called rhabdomyosarcoma. While the causes of most breast cancer and of rhabdomyosarcoma are unknown, known risk factors include genetics; lifestyle activities such as alcohol, tobacco, and drug use; and environmental factors, including certain kinds of radiation, pesticides, and air and water pollution. A 1998 study by the New York State Department of Health linked leukemia and bladder cancer to living near landfills in Suffolk County.[30] But cancer, like radiation, is an emotionally charged issue, and alarming even if the risk is low. Were the incidences statistical fluctuations or real? If real, were they associated with the lab, industries, fertilizer use, or something else? The less one trusted the government agencies that had conducted the studies, the more uncertain were the answers—or the more certain one was that the agency officials were lying.

A third development heightening groundwater sensitivities on Long Island was widespread publicity given to several episodes elsewhere in the country where industries had covered up chemical contamination. One was in Love Canal, New York, where the Hooker Chemical Company had dumped tens of thousands of tons of toxic chemicals; the site became first in line on the Superfund National Priorities List finalized in 1983. In Woburn, Massachusetts, three industries had put chemical wastes in the groundwater, which was thought to be responsible for a high incidence of leukemia and other health effects. After a lawsuit filed by environmental lawyer Jan Schlichtmann, the Woburn site eventually had what was then the largest and costliest ($68 million) chemical cleanup in the Northeast—though Schlichtmann suffered from his efforts, having to declare personal bankruptcy and even sell some of his clothes. The episode was the subject of a 1995 bestseller *A Civil Action*, later made into a major motion picture.[31] A third high-profile site was Toms River, New Jersey, where contaminants from a chemical plant were a likely cause of some childhood cancers. Yet a fourth took place in Hinkley, California, where the Pacific Gas and Electric Company dumped contaminated water into the groundwater in a case investigated by the clerk and then activist Erin Brockovich, eventually resulting (in 1996) in a settlement of $333 million—an episode also turned into a movie, *Erin Brockovich*.

In the 1990s, these three developments—heightened concerns about radiation, cancer clusters, and publicity about chemical contamination—left many Long Islanders with serious questions about matters affecting the health and welfare of their families, questions sometimes with no clear answers. Brookhaven National Laboratory came under particular scrutiny. It was, after all, a Superfund site, and one with nuclear reactors.[32] In 1995, attention to Brookhaven's environmental issues escalated—ironically, thanks to its effort to upgrade the sewage treatment plant. To construct the upgrade required possibly pumping some water slightly contaminated with tritium and other chemicals, at 10 percent of the drinking water standard, into the Peconic River.

Long Island has some dynamic and colorful activists. The sewage treatment plant upgrade drew the attention of Bill Smith, a sport fishing boat captain who had founded a nonprofit organization called Fish Unlimited to focus attention on aquatic environmental issues. Independent and aggressive, Smith—who lived on Shelter Island at the tip of Long Island—was convinced that the lab's pollution had caused fish kills, brown tides, and a lack of shellfish in Peconic Bay, and was determined to oppose any more of what he saw as lab assaults on local waters. He began showing up at public meetings to agitate against the lab's upgrade—and he was an intimidating presence, burly, loud, and swaggering, with a ponytail and facial stubble, and prone to extreme (hence highly quotable) statements. His cause was joined by Karl

Grossman, who wrote a column called "Suffolk Closeup" for some Suffolk newspapers and was the author of *Power Crazy: Is LILCO Turning Shoreham into America's Chernobyl?* Calling BNL a "reckless taxpayer-financed nuclear establishment," Grossman had been campaigning for the lab's closure for nearly two decades. He suggested that what was happening at Brookhaven was like Chernobyl, said that Americans should not tolerate the fact that their tax dollars were supporting a lab that was killing them, quoted a person who described the mentality of BNL scientists as "they'd roll around in a pile of plutonium," and wrote that BNL's history could be entitled "The Years We Poisoned Long Island."[33] The efforts of Smith, Grossman, and others delayed construction of the new sewage treatment plant.

Meanwhile, the continuing Historical Legacy survey (required as part of the lab's Superfund process) turned up VOC contamination in four wells off-site, though how many were a result of lab activities was not clear. The lab was also the target of a variety of rumors that it used tritium for (nonexistent) on-site nuclear weapons, that the BGRR had melted down and the lab had covered it up, that cancer rates were higher in and around the lab, that the lab had poisoned Long Island's water, that cases of the childhood disease rhabdomyosarcoma were higher around the lab, and more.

The lab did not address these concerns effectively. The divided nature of its public engagement activities was one reason, with community relations and outreach separated from media relations. Geiger reported to Gunther of the OER, who reported to Davis, while Rowe reported to the media relations head, who reported to Mark Sakitt in the Director's Office. Geiger's role was to be a "grassroots" advocate, establishing personal relationships, cultivating community trust, going door-to-door, and attending several civic meetings each month to talk about lab cleanup efforts and invite the members to meetings. This activity acquainted her with community fears and concerns, and she tried to devise ways of addressing them. The role of Rowe and the media relations staff's role was less to connect with neighbors than to communicate with the media—to convey information, defend against misinformation, and show off the institution. The media relations staff would, for instance, stress the lab's participation in breast cancer studies on Long Island, and research into cancer diagnosis and treatment. They would produce information about how the amounts of radiation associated with lab work was much less than natural sources of radiation in the environment—sand and soils contain thorium, orange juice and bananas have a radioactive isotope of potassium, and tobacco contains radioactive forms of polonium and lead—and about how human activities such as chest and dental X-rays and high-altitude plane flights also expose humans to radiation.[34] The two roles—community outreach and media relations—each had its strengths and improved over time, but were not coordinated as well as they could have been.

But neither the community relations nor the media relations staff had scientific training or authority. As a result, lab scientists were called into service to support the communications efforts. Unfortunately, when lab scientists addressed the public they would sometimes see it as an opportunity to teach how little risk there was, and come off as implying that any public concerns were groundless and irrational. Brookhaven's health physicist Andrew Hull could be particularly grating, for he was vocal in asserting that the existing safety standards for chemical and radiological exposures were preposterously strict—established for political show rather than scientific substance. Only half in jest, he proposed a "truth in labeling" in which official standards would be designated as "far in excess of the value of whatever reduction in hypothetical radiation risk it may achieve," and as implemented "solely to allay public concern," which might help get "on the path toward restoring reasonableness."[35] Such an attitude left many people feeling that they were not being taken seriously and that the scientists weren't even attempting to understand their concerns and fears.

The DOE's actions inadvertently elevated local fears and anger. Their officials, too, had little training in risk communication and public involvement. When, as a goodwill gesture in light of the VOC contamination, the agency offered free public water hookups to eight hundred homeowners south of the lab, the action backfired and looked like an admission of a major health disaster.[36] Nobody gives away money, least of all the government, so free hookups obviously meant that the DOE agreed that the lab was guilty of poisoning the groundwater, and thus their drinking water. "Digging Up Residents' Fears," one *Newsday* article announced. "Suspicion rather than gratitude for free water hookups."[37]

Prior to the free water hookups announcement, the DOE requested that the lab hold a public meeting to answer questions, to take place on January 16, 1996. The request came between Christmas and New Year's, and while Geiger and the OER staff protested, saying that they needed more time to plan, and that the community had to be involved beforehand, DOE officials insisted they proceed anyway and included the meeting in its press release.

The result was disastrous. "It was the tipping point," Geiger said.[38] In the wake of the ensuing public meeting the lab's relationships with most civic groups eroded or fell apart.

It was the meeting's drama rather than the facts presented there that was responsible for the disaster. The meeting was held at Brookhaven on a gloomy evening in the 450-seat auditorium of Berkner Hall. Grossman, who had a flair for the theatric, came early to the meeting before fire department officials closed the doors after the auditorium reached capacity and directed latecomers to overflow rooms. Standing in the aisle next to where Bond was sitting, Grossman told a television reporter that he was going to stage a scene and asked the reporter to videotape it. Puzzled, the reporter asked why. Because I've been trying to shut this lab for twenty-five years, Grossman answered. He walked out of the auditorium and dramatically tried to reenter the now-locked door, shouting that the lab was barring the press. Rowe, witnessing this, dashed into the auditorium, grabbed a lab employee, begged him to go to an overflow room so she could give Grossman the spot, and returned within seconds. Grossman's theatrics undone, he re-entered the auditorium without incident.

The meeting began with talks by BNL staff on hydrology, the description of the VOC plumes that had migrated off-site, and cleanup methods. Their talks focused on details of the extremely small amounts of pollution involved and how they were below government standards. Karim Rimawi, director of the New York State Division of Environmental Radiation, and an expert with no connection to the lab, spoke last. He gave a detailed presentation of possible health effects in a soft, slightly accented voice, but was shouted down by audience members who accused him of minimizing dangers, covering up lethal threats, and being an outsider unconcerned about the community. The audience was raucous—some wanted more hookups, some were upset about having to now pay for water, some demanded bottled water, some wanted the lab shut down, others accused the lab of killing children, a few cried. The meeting ended with audience members mistrustful and angry, and the lab scientists stunned. Grossman entitled his subsequent column about the event "A Patch of Hell" and described it as taking place on an evening whose weather was "a cloak of radiation-spiked fog."[39] Grossman quoted an activist as saying, "They're [lab employees] the wrongdoers. We're the victims. We're going to get them."[40]

At a Suffolk County legislature Energy and Environment Committee meeting about two weeks later, antinuclear activists distributed a handout proclaiming that "Brookhaven Lab Nukes Are Poisoning Our Air, Our Water, and Our Seas," stating (without backing) that "Suffolk County has the highest breast cancer mortality rate in the nation" and (again without a source) that the highest rate in Suffolk County is in the fifteen miles around the lab. They also circulated a flyer inviting residents to join a $1 billion class-action lawsuit; the Manhattan-based law firm involved would soon also send "Dear Neighbor" letters with the same request.[41] Rowe sent the handouts to lab officials, saying she was letting them know "what is brewing out there," namely, "a movement still in its early stages, but about to cascade into something that may one day match the size of the anti-Shoreham crusade."[42]

Waking up to what was happening, the Suffolk County legislature established a task force to investigate possible health and environmental impacts connected with the lab.[43] Regulatory agencies from the state and county, as well as the EPA, were represented, and meetings were public. The task force was headed by Roger Grimson, an eminent epidemiologist from the nearby Stony Brook University. The New York State Department of Health studied traces of radioactive materials in the Peconic River, finding that the dose was a fraction of what would call for remedial action.[44] That March the DOE funded more public water hookups south of the lab, and then in August about 100 more, saying it was not necessary but would reassure neighbors. The DOE had not learned the clear lesson of its previous action; the new hookups only reinforced the conviction that the lab was poisoning more and more of the environment.[45]

Bill Smith—saying he was adopting a strategy of the Civil Rights movement—called for alumni of the nine AUI universities to boycott their alma maters for "poisoning life on Long Island" until the universities divested themselves of the lab.[46] A website called FABLE (Foundation for Accountable Brookhaven Lab Endeavors) charged the lab with falsifying federal documents, embezzlement, conspiracy, unethical practices, and using Long Islanders as guinea pigs. A disclaimer read, "FABLE makes no warranty for its accuracy."

Throughout the rest of 1996, impassioned conversations about the lab took place all over Suffolk County. A lab biomedical repair technician in a checkout line at a grocery store overheard a cashier and a customer talking about how real estate people were canvassing the neighborhood advising people to sell *now* before outsiders learned of the lab's "cover-up" of its contamination—and how Bill Smith and others were making "almost daily" phone calls to people trying to shut the lab. "I felt sick" at this point, the employee wrote in a letter to a lab environmental scientist. Fortunately, she wrote, the conversation was then joined by someone who felt that the lab was safe and its scientists honest. But the conversation had no resolution.[47]

LONG ISLAND ENVIRONMENTAL ACTIVISM: "A CONTACT SPORT"

The Long Island community was composed of diverse groups whose members had varying attitudes and concerns about the presence of a federally sponsored atomic laboratory in their midst. The tens of thousands of people who attended popular open houses, and schoolchildren who visited on field trips, were generally excited about the lab and its science. Various community groups contacted the lab's Speakers' Bureau to invite lab scientists to address them. But many others in the community either did not know what the lab did or were suspicious of its activities, with some motivated to organize and seek improvements in performance or to demand explanations. Long Island's civic advocacy groups and antinuclear activists therefore had a range of aims and intentions. Some sought a dialogue with the lab over environmental issues, while others simply wanted it closed. In 1996, partly thanks to differing views about the lab, the groups began battling each other. As many of their members liked to say, "On Long Island, environmental activism is a contact sport."

At the beginning of 1996, Jean Mannhaupt, an experienced and self-taught groundwater activist from Manorville, just southeast of the lab, grew concerned about the allegations about Brookhaven's chemical contaminants. Mannhaupt, of the Manorville Taxpayers' Association, was joined by Nanette Essel of the Yaphank Taxpayers & Civic Association, in the town just west of the lab. One of Suffolk County's most active community leaders, Essel was experienced with groundwater and landfill issues, including those at BNL, and in 1997 was the Democratic candidate for town supervisor. Together, Essel and Mannhaupt had a combined grassroots advocacy experience of thirty-eight years and were well known on Long Island. Starting in January 1996, the two worked five days per week, without pay, organizing regular public workshops on issues ranging from environmental laws and regulations to risk assessment. They invited outside speakers, including Paquette, Gunther, and other scientists from the lab, and representatives from the Suffolk County Department of Health Services (SCDHS), the New York State Department of Health, the Department of Environmental Conservation, the Agency for Toxic Substances and Disease Registry (ATSDR), and other key agencies. Mannhaupt and Essel grilled them all, and had trained themselves well enough that they could spot gaps in the lab's soil and monitoring data sheets, surprising the scientists. Said Mannhaupt, in her heavily accented Long Islandese, "I don't think they really believed that anybody in the community could read them."[48]

Mannhaupt and Essel soon found the time and energy overwhelming, and asked Leland Willis, AUI's environmental lawyer, if he could get the DOE to support their efforts, as the agency had done at other national labs. The DOE refused, calling the lab's problems comparatively insignificant. Then, in March 1996, Willis convinced the AUI trustees to award them a $25,000 seed money grant, only to find that, as a nonprofit institution, AUI could not give money to Mannhaupt and Essel directly, but only through another nonprofit. To solve the problem, Mannhaupt and Essel turned for help to Judy Pannullo, director of the Long Island Progressive Coalition (LIPC).

The LIPC had been formed during the Shoreham protests in the campaign against the reactor. It began without staff or resources, but slowly grew to help coordinate the efforts of a dozen or so activist groups pursuing a wide range of class, race, economic, labor, antinuclear, and other issues. LIPC director Pannullo was sometimes called as the "conscience" or "referee" of Long Island's activists. A commanding presence, she had learned to quell fractious meetings by standing up with deliberate slowness—she was tall—hold high a pad of paper, and shout, "What do we need to know and how do we get it?" The LIPC was suspicious of Brookhaven, "its 'cold war' secrecy, its community insensitivity, its environmental degradation, its potential threat to neighborhood health and safety, and its scientific arrogance."[49] Sympathetic to Mannhaupt and Essel's aims, Pannullo used the Research and Education Project of Long Island (REP-LI), a tax-deductible legal entity that could dispense grants—a 501(c)4—to take AUI's money and gave it to what Mannhaupt and Essel called their Community Work Group (CWG).[50] The money didn't go far, not for two people working five days per week for an indefinite number of weeks. But the lab helped by giving the CWG an on-site office, badges, and access to whatever people, material, and other information they wanted.

"We wanted to know: What have you done? Where'd you clean up? What are you doing now?" said Essel. "We had terrific access; they opened the doors, told us everything that was going on."[51]

The CWG was the kind of dialogue and deliberation group between the lab and the community that Brookhaven had needed decades before. "We thought of ourselves as the eyes and ears of the community, but on the inside," Essel said.[52] The group's mission was to

> Provide an opportunity for citizens to become knowledgeable and promote an interactive dialogue with Brookhaven National Laboratory and other interested and involved agencies regarding monitoring and other plans of action which have the potential to impact the community. We expect our participation will result in beneficial changes.
>
> Provide a consistent and well publicized program of public meetings, workshops and other informal avenues for active community participation and education in order to disseminate information on a range of environmental and related issues.

Provide open communication between citizens and governmental agencies regarding federal, state and local requirements with which the Brookhaven National Laboratory must comply.[53]

Community meetings were often rocky confrontations. The attitude of many neighbors was that Brookhaven was a mysterious laboratory sprawling over hundreds of acres and employing thousands of people in research projects they didn't understand or knew little about, which therefore worried them. The attitude of many lab scientists and administrators was that Brookhaven was a world-class, Nobel Prize–winning laboratory doing cutting-edge science in the national interest according to established procedures, so there was no need for concern. The trouble wasn't so much either group's attitude, for each was trying to make sense of the situation given the information they had. The trouble was the existence of the gap between the two attitudes. The need to overcome the gap was obvious—but how to do so, and even what each side needed if it were to happen, was not. The CWG was the only group making a serious attempt.

Furthermore, though Mannhaupt and Essel tried to close the gap, many community members remained suspicious, and came away with a different opinion of the lab's openness than theirs. Despite the open houses, Speakers' Bureau, and educational visits, many neighbors either distrusted, were unimpressed by, or did not know about these.

"The lab operated as a foreign nation," recalled Adrienne Esposito of the Citizens Campaign for the Environment. "No transparency, no communication, no respect. It was a research lab in the middle of a community where nobody knows what they do, and all we know from the press is that the groundwater is contaminated. And then someone from Brookhaven tells us, 'It's fine.' Sorry! We need more than that!" At the community meetings, Esposito added, "I think they thought they were going to convince us and that we were crazy. I think they expected they would go through the motions, give us a dog-and-pony show, pacify us, and allay our concerns instead of responding to them."[54]

The CWG's accepting AUI's support backfired immediately and disastrously. Other activist groups demanded to know why the CWG had gotten the money rather than them, and demanded their own. Groups who thought the sole aim of activism should be to close the lab detested the dialogue that the CWG was fostering, demanded similar access to the lab, and when it was refused accused Essel and Mannhaupt of being tainted, untrustworthy, and corrupt. The CWG now found itself unexpectedly fighting not the lab, but other activist groups.

One of the CWG's most vociferous adversaries was Peter Maniscalco. Maniscalco's scraggy hair and peace pipe gave him the air of a hippie. A truck driver, he doggedly pushed one cause after another for decades, promoting them through hunger strikes, Navajo ceremonies, prayer vigils, lighting candles in the rain, belly dancers, and fire rites. He was an antiestablishment figure from the 1960s, convinced that all mainstream institutions were corrupt. The media cooed over him. "There are people who get attention because . . . there's just something about them," an admiring *Newsday* columnist wrote, "with his beard and his glasses and his red-checkered lumber jacket, and his occasional mischievous smile, and the way he talks about kids—that makes you want to ask where he parks the reindeer."[55] Maniscalco was indeed adorable, so long as you weren't on the other side of one of his causes. In the 1970s his cause had been to close the Shoreham Nuclear Power Plant; in the 1990s it was to close Brookhaven National Laboratory, and he attacked any group whose mission was anything short of that. He would charge that the lab posed an imminent dire threat to the community, that the BGRR had had a core meltdown, and that the HFBR was in danger of having one. At one meeting shortly after the CWG grant was announced, Maniscalco began shouting, preventing others from speaking in what amounted to an impromptu filibuster. "Enough! Let's move on!" said a member of the Affiliated Brookhaven Civic Organizations, but Maniscalco continued to scream.[56] "It was frightening," recalls Essel. "I feared he was going to punch me."[57]

That was only the beginning of savage attacks on the CWG by other activist groups. Bill Smith threatened to charge Essel with slander for some of her remarks.[58] In another bizarre and disturbing episode, a participant at many CWG meetings—whose antinuclear activism was inflamed by having lived near Three Mile Island at the time of its accident—made her way into Mannhaupt and Essel's office when they were absent, convincing Mannhaupt's junior high school daughter to let her in, and made off with some documents. Furious, Mannhaupt and Essel accused the woman of theft and contacted police. "I couldn't believe such things," Essel recalled. "I was on their side! Why were they treating *me* as the enemy?"

Sue Davis wrote an open letter defending the CWG, saying that Mannhaupt and Essel were community advocates, that they had organized and sustained the CWG, and that the AUI grant was not an attempt to co-opt anyone in the community but rather "to engage in open and effective interaction with the Laboratory."[59] Davis also pointed out that the CWG's critics were themselves tainted because they were pushing a class action suit against the lab and wanted to encourage others to join, and had everything to lose from dialogue and cooperation. But CWG's now-enemies easily discounted her letter, coming as it was from a lab employee. Davis now made Geiger report directly to her.

Backlash occurred after the BNL Public Affairs office asked for Fish Unlimited's Form 990, required of all nonprofit organizations. Smith refused, and after he told the CWG that the lab was seeking to intimidate him they demanded that the lab retract its request. "The request was not meant to threaten him, but was instead a straightforward request for information," the office's manager wrote. "On many occasions, Mr. Smith had asked for information about BNL, and the Lab has provided whatever he wished. We have nothing to hide. We don't expect that he has anything to hide." But the office apologized and withdrew the request.[60]

The antinuclear activists' vociferous attacks on the CWG severely damaged its credibility, and antinuclear activist groups whose mission was to close the lab—who were less constructive and promoted less dialogue—began to attract far more attention and media coverage. One was Helen Caldicott, a charismatic antinuclear verbal bomb-thrower. In her autobiography, *A Desperate Passion*, she told of her youthful terror of nuclear war and her "desperate passion" to have a public voice. She had put aside her training as a physician to practice "global preventive medicine" as an antinuclear activist. Her vivid statements and outrageous actions were arresting: she equated working for the defense industry with support for the Holocaust[61] and claimed that "Every time you turn on an electric light, you are making another brainless baby."[62] She insisted that "Every man in North America has plutonium in their testicles as a result of nuclear weapons testing fall-out in the Fifties,"[63] and energized one listless peace conference by organizing a nude demonstration.[64] Actions such as these, as she recounts in *A Desperate Passion*, sometimes landed her in trouble, as on the TV show *Nightline*.[65] In 1984, after over-the-top remarks helped force her resignation as president of the Physicians for Social Responsibility, she wrote that she had been "psychologically raped" by her colleagues.[66] To support her claims, Caldicott often appealed to the work of Jay Gould, a maverick economist turned activist. Caldicott's speeches galvanized audiences. "She's always dynamic and she makes you extremely depressed," said one listener; "I just can't believe this [what Caldicott said about the lab] is going on," said another.[67]

In the fall of 1996 Caldicott gave a talk as part of a service at the Unitarian Universalist Fellowship in Bellport on the evils of Brookhaven. She told the audience that any amount of radiation is harmful, that they should refuse even dental X-rays, and that all Brookhaven employees were protected against the lab's radiation because they had been given special pills. Recognizing that these and other of Caldicott's statements were false, an attendee from the Wading River Civic Association asked Caldicott if she had visited the lab or spoken to people there. "No, and I never will," Caldicott replied. "They [the churchgoers] never sang their last hymn," the distraught questioner wrote Essel, adding, "On the way out I was stopped by a man who accused me of killing his child and people of Long Island. His attitude was so aggressive and nasty, I thought he was going to hit me." Another person confronted the woman to say that she should have stayed home and her husband attended instead. "I left the meeting shaken and upset, that, especially in a church, people would knowingly give misinformation and half truths to frighten the public unnecessarily. . . . Yes, we need to care, be vigilant, and educate ourselves to avoid the mistakes of the past. But, you do it with reason and understanding, not blind fear."[68]

Many lab employees were enraged after activists began approaching churches, recruiting representatives with no understanding of the science to make incorrect and sensational statements. When a member of the General Board of Global Ministries wrote that exposure to even low-dose radiation is a "blasphemous threat to life granted us by the Creator," health physicist

Gary Schroeder pointed out to the Ministries how everything in the natural environment exposes us to doses far more than does the lab, remarking wryly, "You should be aware that the Creator placed us in a world filled with radiation."[69]

After Caldicott gave a talk in the Hamptons at which she urged the audience to go door-to-door in the wealthy community to enlist support to close the lab, an alarmed Rowe wrote the lab management that "you need to be aware of this growing anti-BNL movement in the East End," and sent them newspaper clippings about Caldicott and her activities.[70] Rowe began making a point of showing up at public meetings, even those organized by the militant activists, to present the lab's side. Inevitably, her actions resulted in an "us versus them" atmosphere, which the media then ceased to evaluate but rather reported on the controversy itself, feeding the flames and claims of the hard-core and extreme activists in the process.

Inevitably, too, the interactions between the lab outreach and the militant activists took on personal dimensions. One particular target was Davis, a lab manager and a woman. At one meeting an antinuclear activist shouted at Davis that she was a "cunt"; when she replied with a cool but pointed putdown, it further annoyed the activists and left other lab administrators wishing that Davis had remained silent. Another target was Rowe. Petite, aggressive, and of Hawaiian ancestry, she was reviled by the militant activists as the "Dragon Lady."

In November 1996, the editor of a local newspaper told a reporter from *Nature* magazine that while opposition to Brookhaven does not run very deep, the lab is "quite obsessed" by it, and that "they feel like a great whale with harpoons hanging out of its body."[71]

* * *

The year 1996 had been hard on BNL's media relations staff. They had been insulted, cursed, shouted down, and accused of lying at public meetings. Each December, their office hosted a Christmas party where the tradition was to sing Christmas carols—led (until her retirement) by Sillas, with piano accompaniment by Rowe—with witty lyrics rewritten to reflect lab events. A favorite was "The First Nobel," sung to the tune of "The First Noel."

Someone at that year's Christmas party sought to defuse the stress with a revamped version of "Let It Snow!" called "Let It Go!"[72]

Some neighbors outside are frightful
Tho others can be delightful,
But don't let them get you low
Let it go! Let it go! Let it go!

In another song—written and circulated for laughs but not singing—the group of about twenty exhausted media relations people took advantage of the intimacy of the private occasion to vent their anger about the slander, abuse, and personal insults they had endured.

Jingle Bells, Bill Smith Smells
We hope he goes away
With Jay and Karl and Helen too
In a no-horse close-ed sleigh

Less than a month after that Christmas party, Brookhaven's media relations staff issued a press release that an unexpected amount of radioactive tritium from an unknown source had been found in groundwater at the lab. To many in the activist community, the leak was a sign that the HFBR, and the lab that had built and maintained it, was a deadly threat to Long Island. Among other things, this meant that all the people mocked in the song would be back—with stronger voices, more credibility, and a larger community. The cascade that Rowe feared was underway.

D'AMATO AND FORBES

The January 18, 1997, announcement of the tritium leak was about to affect not just local activists but politicians, in particular New York Senator Alfonse D'Amato and Congressman Michael Forbes.

New York's senior senator was the urbane and affable Daniel Patrick Moynihan, a Democrat raised in New York City known for his principled stances, foreign policy expertise, and mellifluous voice. He would pay little attention to what was happening at Brookhaven over the next year.

D'Amato, a Republican, was his opposite. Raised on Long Island, where the lab was located, D'Amato had a reputation as the go-to person to fix local problems—"constituent services," as they were known—no matter how puny, as long as it brought votes, which is how he managed to remain a Republican senator in a Democratic state for three terms despite a 1991 investigation by the Senate Ethics Committee. D'Amato was a master at the political game; he always had a lively quote and painted himself as the people's protector. Critics dubbed him "Senator Pothole," a nickname he embraced, and "Senator Shakedown," which he didn't.

D'Amato's personal and political behavior were legendary. When he wanted something badly from a peer or heavyweight, he'd nag in his high-pitched bark until he got his way. When angry with subordinates, he'd often "go Godfather," as one congressional staffer put it, calling them into his office, starting gently—"How's the family?"—then launching into a tirade until they were trembling and in tears, before returning to niceties—"Got any plans for the weekend?"

For a long time D'Amato was friendly to and supportive of Brookhaven and got along well with Samios, an equally down-to-earth Brooklynite. "We spoke Newyorkese to each other," Samios recalled. D'Amato helped the lab fight budget cuts, and sometimes acquire new facilities. When D'Amato met Samios the two would greet each other with bear hugs and a loud, "How the fuck are ya!"

In 1997 D'Amato's reelection prospects had dimmed. He had squeaked by his last reelection on a 1 percent margin, and his upcoming reelection campaign in 1998 would be even more difficult. His environmental credentials were horrendous—in 1996, the League of Conservation Voters rated him at 0 percent.[73] Thanks to the anger stirred up over the chemical plumes and the 1996 meeting, the tritium discovery provided D'Amato with a God-given opportunity. Brookhaven National Laboratory would be D'Amato's next pothole.

Michael Forbes, the second-term congressman who represented Suffolk County, was D'Amato's friend and protégé; D'Amato had presided at Forbes's wedding. His first interaction with the lab shortly after winning his election upset him when the lab invited the sitting congressman to an event but not him. However, the relationship improved, and Forbes visited the lab for the first time three months after taking office in 1995, attended the January 1996 community meeting, and attended the groundbreaking of the new waste management facility; lab director Samios took him around, showing him the Relativistic Heavy Ion Collider (RHIC) then under construction. But Forbes soon began to resent the fact that Samios never dropped in on him, despite his position on the powerful Appropriations Committee and the fact that he sought to increase federal dollars going to the lab, one of the largest employers in his district. Like D'Amato, Forbes was on the lookout for environmental credentials.[74] Forbes was also especially sensitive about the lab because he had been vilified by activists for being insufficiently zealous in addressing the lab's chemical contamination.[75]

Also like D'Amato, Forbes was hard on staff. To lighten the office mood, one staff member devised an "f-meter," with the person who had received the most curses that day receiving a shout-out. Forbes was known for flip-flops, saying he wasn't inclined to vote to impeach President Bill Clinton, then changing his mind. A few years later, when it looked like Republicans would lose the next election, Forbes would switch political parties.

The political aspirations of Forbes and D'Amato were another driver of the events of the next year. They were both infuriated that they had to learn about the lab's tritium news from the newspapers, and blasted the lab for incompetence. "They keep drilling around, and that's fine," Forbes said, "but my concern is, what are they doing about getting at the source of the leaks?"[76] D'Amato cited Rowe's failure to mention tritium at her Southampton Rotary Club talk as proof of the lab's duplicity. Both politicians demanded that the reactor be ordered shut until the source was found. But Samios had already done that. The politicians' statements soon grew still more radical and threatening; Forbes said that the tritium leak had "put the future of the lab in jeopardy."[77]

THE DEPARTMENT OF ENERGY

In 1997, the part of the DOE that oversaw most non-weapons national labs was the Office of Energy Research, whose director was Martha Krebs.[78] But several other DOE offices were also involved. Tara O'Toole, as assistant secretary of the Office of Environment, Safety and Health, oversaw those matters, while Terry Lash, as director of the Office of Nuclear Energy, Science and Technology, oversaw the performance of the HFBR. These offices often jockeyed for attention and influence.

Krebs had started out as a theoretical physicist and was an accomplished science administrator who had spent six years as a staff member for the House Committee on Science and then ten years as an administrator at the Lawrence Berkeley National Laboratory before joining the Office of Energy Research in 1993. Her appointment was confirmed by the Senate a month after Congress canceled the Superconducting Supercollider (SSC), and her first task was to stabilize the country's high-energy physics program in its wake. She was experienced, cool-headed, and had a reputation for running a hard-working office, though she was not experienced in being in the dead-center of political controversies.

O'Toole would be the principal force driving the DOE's response to Brookhaven in the next three months. Trained as a physician, she had spent five years as a senior analyst at the Congressional Office of Technology Assessment before President Clinton nominated her for the DOE position in 1993. Conservative senators held up her confirmation after they discovered that, while doing her residency at Yale, she had joined a reading group called "Marxist Feminist Group 1." Hazel O'Leary, the DOE's first Black female secretary, had to mount an extensive campaign to get O'Toole confirmed.

Lash also had trouble at his confirmation hearings. He had been the staff scientist at an environmental group before becoming head of the Illinois Department of Nuclear Safety in 1984. He resigned in 1990 after accusations that he had misled the state legislature about whether there was an aquifer beneath a proposed nuclear storage site and had altered draft documents to remove the word "aquifer." He nevertheless was confirmed in 1994 as head of the Office of Nuclear Energy. He was a major promoter of a program to improve nuclear safety in Russia and the Ukraine, called the "Lisbon Initiative" after the site of a key meeting to establish the program.

By this time the DOE was under heavy pressure to focus more intensely on environmental issues at its facilities. While the AEC had largely protected national labs, even weapons labs, from environmental laws, the DOE could not. One key step was the 1976 Resource Conservation and Recovery Act (RCRA), a federal law governing disposal of hazardous wastes, as well as a 1984 court case (*Legal Environmental Assistance Foundation v. Hodel*) requiring the DOE to comply. These developments significantly changed the DOE's attention to environmental issues at its labs. One result was the FBI raid of the Rocky Flats nuclear weapons production plant near Denver, Colorado, which resulted in the closure of the plant and in its contractor, Rockwell International, being charged with environmental crimes. In the early 1990s, DOE Secretary Admiral Watkins sent teams of inspectors to all the labs to identify environmental and safety issues; Watkins called them "Tiger Teams" from his navy experience where similar groups operated. The Tiger Teams would swoop into national laboratories, initially weapons labs but eventually all DOE labs, to check for violations of best practices, quality assurance, and environmental law. Many labs protested vociferously against what they thought were the extreme lengths that the Tiger Teams were going to, such as citing violations for paint brushes being left to dry before being discarded, or oscilloscopes without the proper labels. But the Tiger Teams did turn up some more serious issues. The final reports of many of these teams often cited the labs for putting science above safety.[79]

Until 1997, O'Toole had focused on environmental problems at weapons production plants. By comparison, Brookhaven's problems were small—but this didn't matter, for two reasons. One was that she and others in the DOE were annoyed that Brookhaven had been unaware of the tritium. The second was that politicians were now on the DOE's case, for D'Amato had begun calling her up "yelling and screaming," in her words, to pressure the DOE into action.[80]

D'Amato's involvement was difficult for the agency; O'Leary had come under fire for travel junkets and mismanaging expenses, and had resigned in January. To replace her, President Clinton chose then-Secretary of Transportation Federico Peña, but Peña's confirmation hearings were not scheduled until March, and DOE officials were terrified that D'Amato would hold up the hearings. So when DOE officials ordered O'Toole and Lash to "get the hell" up to Brookhaven, as O'Toole put it, and "get it under control because we need to get Peña confirmed," she brought to the task the same determination that she had brought to policing environmental compliance at the weapons labs.[81]

ASSOCIATED UNIVERSITIES, INC.

When Brookhaven and other national laboratories were established soon after World War II, they were managed by contractors (mainly nonprofit) that were buffers between the federal bureaucracy on the one side (initially the AEC, later the DOE), and the labs on the other. This was a continuation of the model that had been set up to run the Manhattan Project during World War II. Each was operated with federal funding yet maintained an independent, university-like research atmosphere. These contractors continued to manage the labs until either they or the government grew unhappy enough to want out. This arrangement was known as GOCO, for government owned, contractor operated.[82]

At its founding, AUI was Brookhaven's contractor. Though founded by nine universities—Columbia, Cornell, Harvard, Johns Hopkins, the Massachusetts Institute of Technology, Princeton, the University of Pennsylvania, the University of Rochester, and Yale—in accordance with its bylaws AUI was always an independent nonprofit organization. Originally, these universities each had an opportunity to nominate one person to the board, but AUI still had to approve them.

Initially, AUI and the lab paid considerable care to the environment and the community. A 1948 memo from lab environmental scientist Bernard Manowitz evaluated the environmental challenges: "1) The laboratory is located in a highly populated area [but far less than in 1997], 2) The water table in this area is high and is used extensively for well water and 3) The nearest streams all pass through populous communities before discharging into the bay." The primary focus of Manowitz's memo was to evaluate methods to dispose of radioactive products. He assumed that low-level wastes at the same or less than the natural level of radioactivity—about 1,000 picocuries per liter—could be disposed of routinely into drains. But, Manowitz cautioned, "Radioactive pollution does not have to be injurious to health to be socially undesirable."[83]

Manowitz and other lab environmental scientists worked with the US Geological Survey (USGS) to study Long Island's groundwater flow. They installed instruments to monitor radiation levels in the air and groundwater, and a test center to analyze samples. Lab scientists researched other environmental issues involving meteorology, oceanography, biology, and botany.[84] AUI also sought to cultivate community relations: "A principal objective of the public education program of the corporation is to maintain a satisfactory relationship between the Laboratory and its neighbors in the surrounding communities on Long Island. . . . A laboratory devoted to the nuclear sciences might all too easily become an object of fear and suspicion."[85] Much of that program was educational, involved discussing the hazards of radiation, which is naturally present in the environment but arouses particular fears because of its invisibility, long-term and difficult to detect effects, and association with the atomic bombs dropped on Japan. Right after World War II, trust in science was strong, and a scientist need only have instruments and explanations to be convincing. One of the lab's most colorful was Willy Higinbotham, an instrument-building physicist who invented the first video game—an early version of "Pong"—as an exhibit for a Visitors' Day. (For his leftist views Higinbotham was also a target of the right-wing demagogue Joseph McCarthy.) Higinbotham once took a Geiger counter to the home of a woman who claimed her ducks were radioactive to prove to her that her ducks were fine.

Over time, however, the initial focus on education and community relations at the AUI corporate level diminished, even as local concerns about the environment rose. More than Geiger counters, charisma, and lab open houses were now needed on Long Island if one wanted to reassure its inhabitants about the safety of a nuclear facility.

Meanwhile, the scale of the lab the AUI was managing grew. Brookhaven's first large particle accelerator, the Cosmotron, began operating in 1952 in a large room inside a laboratory building; the Cosmotron's successor, the Alternating Gradient Synchrotron (AGS), built in 1960, consisted of a ring, half a mile in circumference, that was part of the landscape; and in the 1970s the lab began building an accelerator, at first called ISABELLE and eventually transformed into the Relativistic Heavy Ion Collider (RHIC), 2.4 miles in circumference, so large that it could be seen from space. The Brookhaven

Graphite Research Reactor (BGRR), completed in 1950 and the first reactor built for nonmilitary research, was replaced in 1965 by the High Flux Beam Reactor (HFBR), which the lab was seeking to upgrade in the 1990s. The National Synchrotron Light Source (NSLS), operational in 1982, produced X-ray and ultraviolet light beams and was one of the most productive scientific instruments ever for the scope and number of scientific papers that its research supported.

By 1997, its half-century anniversary, AUI had shepherded the lab through one of the most glorious periods of US science. AUI's president was Robert Hughes, a chemist from Cornell who had served as director of the Cornell Materials Science Center and at posts in the National Science Foundation before becoming AUI's president in 1980. During Hughes's 17-year tenure, the lab's research had received two of the four Nobel Prizes. Under Hughes AUI was not especially politically active, and while it did a serious and thoughtful job evaluating science programs, it had not done the same for nonscience issues. There seemed no need.

But times were changing. On a visit to the lab in the mid-1980s, Al Trivelpiece, one of the predecessors of Krebs as head of the Office of Energy Research, warned the lab's senior management that, after the Tiger Teams, the AEC's culture of partnership at the lab was being replaced by a new, more process-oriented culture at the DOE and that the lab needed to adjust or face consequences. In the mid-1990s, friction developed between the DOE and AUI when AUI was asked to lead Lash's Lisbon initiative to improve nuclear safety in Russia and the Ukraine. Although initially interested, AUI declined after the DOE refused to cover liability. Nevertheless, the DOE continued to rate AUI's performance highly at the contract renewal in 1995.[86]

AUI was aware of the national concerns on environmental issues and moved to address them and improve community relations. At the 1996 October board meeting Davis summarized the lab's outreach, including the CWG and efforts to involve local communities through door-to-door notification and discussion with homeowners, one-on-one briefings with elected officials, and smaller off-site public meetings. The lab's quarterly newsletter reviewing Superfund activities was distributed both on- and off-site. Significant progress, she said, has been made in getting communities around the site involved in reviewing and discussing the issues. However, she added ominously, there remains a small fringe group of activists that is very "anti-nuclear."

Unfortunately, while the activities undertaken by the lab were made clear to the local DOE office, AUI failed to trumpet its own outreach actions loudly to the DOE, whose officials had the impression that the AUI was doing little, and that what it did was not worth supporting.

AUI's trustees were distinguished scientists from a time when scientists had both political and social clout. They included two Nobel Prize winners, including one, Val Fitch, who had won his for work at the lab. They did not fully recognize the new politics and power dynamics in the DOE, where politicians no longer automatically deferred to scientists, not even Nobel laureates, and the public had a greater voice in issues involving science policy. It was not that politicians and people had become less scientifically literate, or even less impressed by science, but rather that the social and political climate had changed—especially with the end of the Cold War and therefore the end of the insulating of the weapons laboratories and other scientific facilities from intense environmental scrutiny—and along with it the authority of those who spoke in its name. AUI was now operating in a far more complex political environment than when the organization was created, and a slight jostle—even a tritium leak of no health or environmental hazard—was all it would take to cause a breakdown.

When the AUI Board met at the end of January 1997 at its Washington, D.C., office, one item involved selecting Hughes's successor in the face of his coming retirement as president, and the board formally approved an offer to Lyle Schwartz, the director of the Materials Science & Engineering Laboratory of the National Institute of Standards & Technology. Another item was the tritium discovery. Samios reported that the lab had good relations with Lash and O'Toole and that they were handling the "tritium situation in a very calm and professional manner."[87]

"Nobody took it seriously," trustee Barry Cooperman recalled. "We had a good reactor doing good science, and the amount of tritium that had leaked was the equivalent of that in an exit sign. We thought we would win out because we had science on our side. It was nothing to lose sleep over."[88]

2 SACKED

We're gonna close the lab down if we have to, to get this done.

—TARA O'TOOLE

CHARACTERIZING THE PLUME

Returning, now, to the leak itself: By February 1997 Paquette's first set of monitoring wells established the rough outlines of the tritium leak. It started somewhere near or under the reactor building and broadened and diluted in a plume that moved south with the groundwater. Paquette's second set delineated its varying concentrations, which ranged from fractions of the drinking water standard to thirty-two times over. He ordered a third round of wells 1,100 and 1,900 feet farther south. No trace of tritium appeared near the lab's border. The lab briefed regulatory agencies daily and issued weekly press releases, each delayed several days by their trips up the DOE approval ladder.

How, then, to deal with the tritium? Gunther, Sue Davis, and Assistant Director for Facilities and Operations Michael Bebon met with others on February 7 to discuss options. Required to follow Superfund procedures, they considered five. The first two options would try to stop the flow entirely, but were technically challenging, hugely expensive, and possibly futile. Two other options would pump the water out at the plume's leading edge or where the tritium concentration was highest. One would put the water in drums, but it would generate huge volumes of water with no easy way to store, treat, or dispose of it. The other would pump water from the leading edge and pipe it upstream, where it would be put back in the groundwater to dilute and decay further in its trip back down. A fifth option was to do nothing; after all, the plume was not a health hazard and would not reach the lab border. But this would mean the "public perception of no action."[1] Though the Superfund evaluation rules mandated considering this option, it was socially and politically unacceptable, as it would certainly outrage Forbes and cause D'Amato to "go Godfather." The best choice, Gunther, Bebon, and Davis decided, was to extract water at the leading edge and pipe it back upstream.

A week later, on February 14, they and area DOE officials traveled to the DOE's headquarters in the Forrestal building in Washington to propose the plan. They discovered that O'Toole and Lash had lost all confidence in the lab, and accused them of incompetence so ferociously that Davis later called it the "St. Valentine's Day Massacre."[2] Lash, as head of the office with oversight of the reactor, refused to believe that it was the source of the plume and asked Gunther about data from the wells west of the reactor. Gunther stared blankly: "What wells?" Lash said he had ordered them and accused the Brookhaven scientists of being "resistive."[3] Confused, the group said they had no recollection of that, but would look into it. Over the objections of the BNL scientists, Lash also said he was bringing in experts from a DOE facility in Idaho to manage the remediation project.

On the way back to Brookhaven, the DOE area office officials sheepishly apologized for Lash's remarks. A few days later, the Idaho team arrived. Unfamiliar with the lab, they did little more than write up a final report—but as Gunther said, "Having them there helped the DOE feel like we were doing the right thing."[4] The DOE wanted not only the perception of action but also the perception of supervision.

Paquette's data showed that it was pointless to install wells east and west of the reactor but the OER staff had them drilled anyway, and the samples came up clean, as expected. Drilling through the building's containment shield for further confirmation was impossible if the reactor were to restart, which lab and DOE officials fully expected it to do. The DOE then proposed drilling wells horizontally under the building using technology created for oil exploration.[5] Meanwhile, Bebon, who had worked for the DOE area office before joining the lab, created a paper trail for every DOE instruction to avoid future accusations of insubordination.

* * *

The lab's finances and research were suffering. Each week more wells were drilled, cleaned, calibrated, and sampled. With dozens of yellow-painted wells already in the ground and more installed daily, the joke circulated that the lab was beginning to look like Swiss cheese. Thousands of samples had to be tracked and analyzed, and an entire building was taken over to store them. Lab employees stopped other tasks, such as analyses of air monitoring and the required urine samples for security personnel, and worked on weekends to analyze groundwater samples. Still, the volume was overwhelming and the lab had to farm out analyses to outside testing centers.

Nine drilling rigs—all those available on Long Island—were putting in ten to twenty temporary vertical-profile wells per week at $5,000–$10,000 each; about fifty permanent wells at $25,000–$30,000 each were installed for monitoring system. Each sample analysis cost $100–$150, and around one hundred samples were taken per day. Each pump and treat construction project cost $900,000, several were required, and the piping alone cost a million dollars. All taxpayer money, going to what lab scientists, DOE, EPA, and county officials knew was not a health hazard. At one meeting O'Toole told Davis and others, "Just get on with the work, I will find the money!"[6] Some money materialized; by late summer about 60 percent of the cost came from the lab's budget and 40 percent from new DOE funding.

Teams of oversight personnel descended from the DOE, EPA, and Suffolk County, each needing to be brought up to speed and briefed daily. Dozens of newspaper, magazine, and TV reporters demanded documents, interviews, information, maps, dates of previous HFBR shutdowns, names of users, tours, photo ops, and more. A reporter from *Newsday* wanted all data from all wells, a *New York Times* reporter wanted all reports and press releases. Some reporters became annoyed and suspicious if they weren't instantly given what they wanted and accused the lab of covering up. Demonstrators began appearing on the road just before the lab entrance with signs accusing the

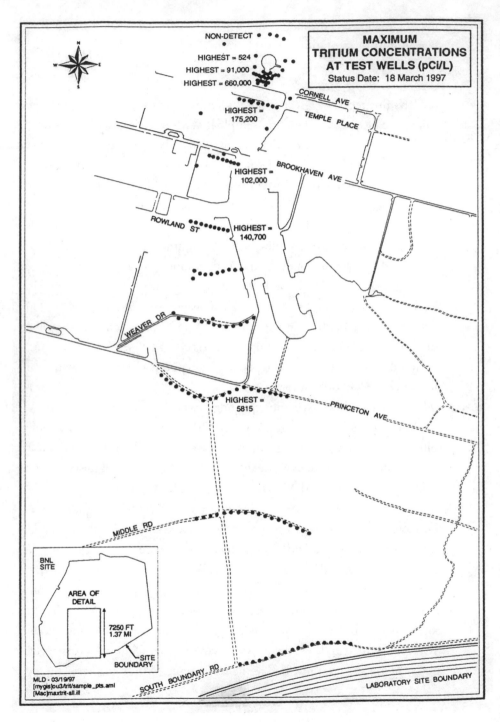

Figure 2.1
Map of monitoring wells planned by Paquette and installed south of the HFBR.

lab of killing children and demanding its closure. Paquette's boss Robert Miltenberger proposed putting up a sign there saying, "LAST YEAR, 3 PEOPLE DIED FROM CAR ACCIDENTS AT THIS SPOT. LET'S FIX REAL RISKS NOT IMAGINARY ONES!"[7] It would have been a sign as much about different perspectives on risks as about risk statistics.

GRANDSTANDING

The announcement of the tritium leak invigorated a spectrum of concerned citizens and groups, from civic associations to antinuclear activists. "People felt like they had been lied to again," said the CWG's Mannhaupt.[8] More militant activists were more strident. Grossman called tritium a "canary," heralding the imminent appearance of more radioactive pollutants.[9] Bill Smith issued a press release declaring that "these bastards are killing people on Long Island and they're getting away with it."[10] Smith told another reporter, "This is the beginning of the end for those reactors, and it's long overdue. These things are an abscess on the surface of Long Island."[11] Maniscalco, announced a "Lenten Prayer Vigil" at the lab entrance to protest its "crucifixion of Mother Earth," to be held on the seven Fridays of Lent from February 14 to March 28, with a special Good Friday service. "As the children of the Earth," Maniscalco declared in a press release, "we just begin to see Mother Earth as sacred and to defend the Earth as if our lives depend on it."[12]

Public institutions are expected to be more transparent, vigilant, and accountable than others, and in the eyes of many neighbors, Paquette's discovery seemed to have demonstrated that the lab had been failing at all three for years. Long Islanders whose knowledge of the lab came from headlines began to see it not as a scientific facility but as a polluter. The Cub Scouts canceled an event at the lab.[13] One mother refused to let her daughter take a bath or brush her teeth with tap water;[14] another refused to allow her child to go to school and said that school buses coming from the lab should be checked for radioactivity. A *New York Times* reporter told a Columbia University physicist working at the AGS that he was an irresponsible scientist for occasionally bringing his children to the lab, for it demonstrated that science was more important to him than his children. "Permanent Shutdown of BNL Sought," proclaimed the cover of *Suffolk Life*.[15]

While employees often encountered supporting voices in the community, they also often encountered virulent reactions. Rowe's daughter, a high school junior, was hounded to sign a petition to close the lab—and a teacher harassed her when she didn't, weaving the HFBR and the radiation leak into daily lessons (segueing, for instance, from the Vietnam War to Jane Fonda to Fonda's appearance in the movie *The China Syndrome* to Brookhaven's reactor). Rowe wrote a letter to the school superintendent offering to address the class about the lab's environmental issues, but was refused.[16] As Rowe walked her daughter to the school bus stop one day, her daughter said quietly, "I wish you had a regular job. Then I could just worry about stuff like the junior prom coming up, who I'm going with, and what to wear."

Brookhaven's scientists were realizing that whenever they tried to communicate facts about the tritium leak, and to make comparisons to the amount of radioactivity in wristwatches (25 billion picocuries—which sounded like a lot, but that was a function of the units used), exit signs, bananas, and ordinary life, it had no effect on activists. Whatever reassurances the scientists provided were coming off as condescending and uncaring.

Rebuilding trust was next to impossible in such a polarized environment. The scientists' habit of treating the environmental consequences as a purely scientific issue—a matter of numbers—did not help. Nor did telling people they were foolish to be upset about it. "They'd say that there's nothing wrong with tritium," Smith said, "it's in exit signs, it's in your watches, it's very safe. Then they'd find more of everything from tritium to plutonium. And then they'd focus on what a great asset the lab was. But you don't drink exit signs or watches."[17] Smith organized a display of banners over Long Island Expressway overpasses on the way to the lab that read, "Entering Nuclear Reactor Zone," and "Brookhaven Lab Poisons Our Air and Water."[18]

Lab scientist Stephen Dewey, who was developing treatments for substance abuse—and had patented one—wound up sitting near Forbes on a flight back from Washington, and politely asked him to explain his statements about the lab; Forbes snarled back that Samios was a liar. "It's one thing if my neighbor calls my boss a liar," recalled Dewey. "But my congressman?"[19] Dewey was so upset that he called Samios, who took it lightly, saying, "That's politics." Dewey, who had been doing outreach to schools about drug abuse for years, now found that teachers and parents were showing up to his talks for the express purpose of shouting accusations that he was causing cancer. "It got ugly," Dewey recalled. He wrote Rowe:

> These people are scared and all the education in the world will have little impact. In my opinion, we need to show more compassion for these problems and we can not make light by making comparisons that involve conscious choices people can make. A scientist in our own department told us that the amount of radioactivity released in this spill was the equivalent of that found in approximately 100 wrist watches. Interesting indeed but this just doesn't cut it! I am not a public relations person but I can tell you that this type of information is meaningless to the people I have come in contact with. When you don't trust someone, or you don't trust a group of individuals, then their words fall on deaf ears. . . . [T]he negativity associated with BNL is growing and growing. People are more interested in my answering contamination questions than they are about my answering questions about our research in drug abuse. . . . Even I am now getting quite concerned about how we (BNL) are handling these very real fears.[20]

Scientists elsewhere in the country were growing worried about the impact of the HFBR shutdown on scientific research. The newsletter *New Technology Week* worried that Long Island's antinuclear activists had generated news stories that might lead to "an irrational, political solution: permanently closing HFBR." This, the newsletter continued, would damage the ability of American companies that depend on neutron scattering, such as Exxon, DuPont, and Genentech, "to create new materials, devise new pharmaceuticals, and understand the workings of matter," not to mention university research and the training of PhD students.[21]

Brookhaven's scientists were relearning the truth of Manowitz's observation—that radiation does not have to be injurious to be socially unacceptable—the hard way, and too late.

* * *

In mid-February, O'Toole and Lash visited Long Island and, pushed by D'Amato and Forbes, arranged still more public water hookups for homes

Figure 2.2
Left to right: Senator Alfonse D'Amato, DOE Assistant Secretary for Environment, Safety and Health Tara O'Toole, and Congressman Michael Forbes at press conference, February 20, 1997. From Newsday. © 1997 Newsday. All rights reserved. Used under license.

in Manorville, at a cost of $6.2 million.[22] Recalling events of the previous year, lab officials begged them to hold off. If the DOE simply announced the hookups it would seem, BNL officials warned, like "this is a major problem for people off site and the DOE is doing this to cover up a real serious health problem . . . that everyone is at high risk" and about to be poisoned.[23] The DOE's Washington office ignored the pleas and proceeded anyway, letting D'Amato and Forbes take credit.[24]

The two DOE officials arrived on February 19, and their first stop was to see D'Amato. The next morning they announced a three-step program to fix the HFBR's problems. In thirty days the DOE would start planning to ship the pool's spent fuel rods off-site; in sixty they would start remediating the plume; in ninety they would schedule installation of a stainless steel lining in the pool. The first happened on schedule, while the second was damaging and premature. It was damaging because it suggested, falsely, that the leak was a genuine health threat, and premature because, when two months later members of the DOE's own expert review panel suggested that remediation was a waste of money, the agency had left itself no room to dial back. The third step occurred, not ninety days, but three years later, after the HFBR had been terminated.

The catchily titled 30-60-90 plan was intended to demonstrate that the DOE had matters in hand. O'Toole said, "The matter of tritium in groundwater at BNL is getting very high attention at DOE." Lash said the HFBR was an important scientific resource and promised that it would not be shut down. "We are committed to taking steps necessary to be sure it will be able to restart in a safe manner." He added, "This is one of the best [research] reactors in the country."[25] But O'Toole made clear to lab officials that they had to take orders from the DOE, with words to the following effect: "We're gonna close the lab down if we have to."[26]

That afternoon, O'Toole and Lash attended a public meeting in Riverhead, about half an hour east of the lab. D'Amato and Forbes said how passionately concerned they were about the environment, and denounced the lab for its "cavalier" and "public-be-damned" attitude. D'Amato charged that the lab had tried to hide the news about the plume by releasing it on a holiday weekend. He praised O'Toole profusely, calling her a "very special person" and referring to her as "the doctor," while O'Toole applauded D'Amato.[27] It was an extraordinary sight: a conservative Republican senator with zero environmental credentials, and the former member of "Marxist Feminist Group 1"—and now the assistant secretary of the Office of Environment, Safety, and Health at the DOE—publicly singing each other's praises.[28]

That public meeting was a turning point in the political dynamics. Before, many scientists could still hope that some local politician would go through the numbers, consult relevant experts, and realize that the lab posed no health hazard and that the proposed remediation was a waste of money. Not now. Few would stand up to D'Amato and Forbes, or question the media's depiction of events.

At the meeting, O'Toole mentioned the changing attitudes about environmental legislation that had taken place over the past half-century. "So if we knew what we know now in 1950, I think we would have been much more stringent in terms of how we built the reactor and how we monitored it." She was then asked, "Would you have built one over a sole-source aquifer on Long Island?" O'Toole responded, "No."[29]

That remark was ominous. For many people, from then on, any evidence that the HFBR was safe, or the tritium plume not a health risk, was beside the point—the reactor should not be there, period, and O'Toole had said so herself. The HFBR had ceased being a facility whose fate was to be decided by evidence and evaluation. Instead, it had become a symbol. To the activists it was a symbol of the dangers of reactors, and to the DOE of the failures of all national labs to care for the environment; meanwhile, Forbes and D'Amato used their opposition to the HFBR as a symbol of their environmental consciousness.

THE HFBR

Neutron beams are essential tools in many scientific fields, and the High Flux Beam Reactor was one of the country's most important scientific facilities, its best neutron source at a time when US neutron sources were heavily overburdened and European research centers were taking the lead.[30]

In 1997, the HFBR was just over thirty years old. Shortly after the Brookhaven Graphite Research Reactor (BGRR) was completed in 1950, lab scientists began planning a more advanced reactor, but the proposal languished in Washington. Then, in 1957, the Soviet Union launched Sputnik, the world's first artificial satellite. When a panicked US government began pouring money into the science budget, one beneficiary was the HFBR, designed for producing neutron beams, along with a sister reactor, the High Flux Isotope Reactor (HFIR) at Oak Ridge, designed mainly to produce californium, an isotope with numerous practical applications. The HFBR had an innovative design that made it more compact and effective at producing neutron beams and was created by five Brookhaven scientists, including physical chemist Julius Hastings, who received a patent for the invention.

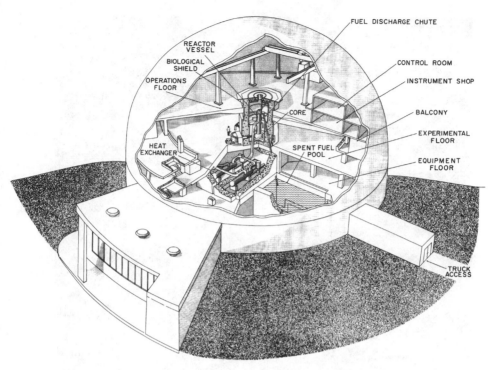

Figure 2.3

Cutaway of the High Flux Beam Reactor showing the location of spent fuel pool.

Its construction began in the spring of 1962 and it became operational on October 31, 1965. It cost $12.5 million and in 1966 reached its design power of 40 megawatts, though it was pulled back to 30 megawatts (nuclear power plants typically are fifteen to forty times that).[31] Supported by the US Department of Energy, the HFBR had an important and clear-cut role as the only high-flux American reactor ever built specifically designed for research that required neutron beams.[32]

The HFBR, the best place in the United States for neutron scattering experiments—rivaled worldwide only by the Institut Laue-Langevin near Grenoble, France—was a vibrant place to work, attracting a diverse group of first-rank scientists. Researchers studied metal alloys, ceramics, magnets, polymers, and other materials used in batteries, solar energy devices, and oil spill recovery. Others investigated protein structure and structures of other biological matter, as well as the mechanisms of Lyme disease, flu-like viruses, and

Figure 2.4
High Flux Beam Reactor core; the fuel elements are in the "bulb."

other diseases. Still other researchers used the HFBR to study nuclear physics, medical applications such as anticlotting drugs, treatments of bone cancer, breast cancer, rheumatoid arthritis, and diagnosing heart disease. Archaeologists used the HFBR to determine the origin of ancient artifacts and trade routes.[33] The HFBR was an important source of medical radioisotopes, including tin-117, used for treating metastatic bone pain; palladium-103, used for treating prostate cancer; three different radioisotopes for treating cardiovascular disease; and four different radioisotopes for treating cancer tumors.

"Experimentalists saw the HFBR as the place to go," said William D. Magwood IV, who had recently joined the DOE's Office of Nuclear Energy, Science and Technology under Lash. Magwood had visited the HFBR several times to conduct detailed studies of its capabilities and of its role in the US isotope production system. "If you couldn't get your experiments done at the

Figure 2.5
High Flux Beam Reactor, experimental floor.

HFBR for one reason or another, then HFIR or NIST [National Institute of Standards and Technology, which had a smaller reactor]would be your second and third alternative. We were proud of our association with it."[34]

The lab took pride in the HFBR and featured it on public tours. When it was offline, visitors could go to the experimental hall to see beam lines and instruments, and the control room. Over thirty years several thousand visitors saw pictures of the reactor's construction, and most were surprised at how small the core was: about six feet across, in the shape of a thimble. The public accessibility of the HFBR reflected and reinforced the comfort of many visitors about the presence of a nuclear reactor on Long Island. So did the fact that the HFBR had an excellent safety record.[35]

When asked for specifics about what the HFBR did, guides found that visitors were generally less receptive to details of nuclear structure and the origin of magnetism than tangible things like the production of isotopes for medical diagnosis and imaging, and improvements to cellphone structure and automobile parts. The down side of such an emphasis on practical applications, it would turn out, was that it made some people think that's the only thing the facility did.

Over the years, the HFBR grew ever more important in the US science program as its instrumentation evolved—including "cold" neutron sources whose low energies were important for biological research—and as neutron sources grew in importance while no new facilities were being constructed in the United States.[36] A crisis was brewing as to how to maintain US neutron research, for its neutron sources were over-subscribed and under-supported. An influential report sent to Krebs said that the HFBR "must be upgraded to obtain essential neutron scattering facilities for the nation's research community" as soon as possible.[37] The HFBR managers developed plans for an upgrade, but it went unfunded year after year.

It was indeed a propitious time for an upgrade, which would allow US researchers to exploit new methods of neutron scattering. Upgrading the HFBR would yield more of the low-energy neutrons that were proving indispensable throughout most branches of science, and would secure the country's ability to produce radioisotopes for medical research and treatment. As the chair of the HFBR's Program Advisory Committee wrote Krebs, "In summary, the HFBR has had an illustrious 30 year history. That impressive performance need not be eclipsed. With relatively modest modifications, the HFBR can be dramatically improved, enough to rival any of the world's best neutron sources for the foreseeable future. Upgrading the HFBR would be a wise investment for DOE and would make a major contribution to American Science."[38]

The neutrons used in reactor research are a byproduct of nuclear fission in the uranium fuel elements of the reactor's core. These fission neutrons are too fast to continue to cause uranium to fission, and the HFBR used heavy water—water whose hydrogen atoms had a proton and neutron in their nuclei—to slow or "moderate" the neutrons. The bulk of the neutrons inside the HFBR were carried off by nine beam pipes to research instruments in its experimental hall. But a small number returned to the core to sustain the reaction, and a few would interact with nuclei of the heavy water to form tritium.

Tritium accumulated in the 68,000-gallon pool of water where the used-up, or spent, fuel rods were stored before their shipment off-site for reprocessing at Savannah River Lab in Georgia. The major tritium source came from the monthly refuelings where the vessel head was removed and fuel was shuffled and replaced as necessary. There were other contributors, like evaporative small leaks from valves and maintenance activities and from the spent fuel that was transferred to the pool several times per year. Molecular exchange between the air vapor and the pool where the spent fuel elements and other radioactive pieces of equipment were stored caused the tritium level in the pool to rise until they equalized.[39] The pool ultimately contained a little over 30 curies of tritium.

This spent fuel pool was 8 feet wide, 43 feet long, between 20 and 30 feet deep, and, like the foundation of the rest of the building, consisted of three-foot-thick reinforced concrete. The walls were lined with a thin membrane and 6" × 9" tiles whose joints were filled with Portland cement, "waterproofed," according to the specifications, "with 10% hydrated lime." The pool's designers knew it would lose around 50–100 gallons of water per day due to evaporation, depending on the pool's temperature. The original design therefore included an instrument to trigger an alarm when the level lowered by a certain amount to signal that the pool needed topping off.[40]

The HFBR, and the employees and users who worked there, were exactingly monitored to detect the presence of even tiny amounts of radioactivity

Figure 2.6
High Flux Beam Reactor, spent fuel pool, full.

entering or leaving the facility. One employee set off alarms when returning to work after a thallium stress test. When another got a few particles of a radioactive substance on his clothing and monitors picked it up, he was required to remove his pants, have them cleaned, and to put them back on before allowed to leave.

The only part of the HFBR that was not exactingly monitored, it turned out, was the spent fuel pool.

PROMISE DELAYED

A Suffolk County environmental regulation enacted in the 1980s had set in motion the events that culminated in Paquette's discovery of the leak in the spent fuel pool.

Brookhaven, like the other national labs, was technically exempt from all but federal regulations. But by the 1980s, state and local officials and politicians were increasingly demanding that federal institutions comply with theirs. The DOE was also becoming highly concerned about environmental issues, one reason being the Resource Conservation and Recovery Act (RCRA), a federal law governing disposal of hazardous wastes, which in 1984 was interpreted as applying to DOE facilities.

In 1987, before the Superfund designation, representatives of DOE, BNL, and the Suffolk County Department of Health Services (SCDHS) signed an agreement in which the lab promised to conform to all local, state, and federal regulations concerning public health and environmental protection, and to make upgrades as expeditiously as possible.[41] Legally, this commitment was unnecessary, as the county and state had no official jurisdiction over the lab. In practice, the agreement appeared to benefit the lab, as a signal that, in environmental health and safety, it was progressive compared with some other DOE facilities. Owners of gasoline stations on Long Island were now expected to protect the groundwater by double-lining their gasoline storage tanks—why not BNL's? "We wanted to treat BNL no differently from other industries," recalls Joe Baier, the SCDHS director at the time.[42] The sea change in environmental expectations, begun in the 1970s, was beginning to collide with the traditional practices of the federal lab system.

Suffolk County's regulations included Article 12 of its Sanitary Code, which required underground storage tanks of hazardous material to have a double lining, though pools could be single-lined. In which category was the HFBR's spent fuel pool?

When built in 1964, long before Article 12, the spent fuel pool conformed to existing county regulations. Leaks of a few gallons per day would not have violated existing regulations, at least not until 1976, when the EPA applied a tritium drinking water standard of 20,000 picocuries per liter.[43] When filled with 68,000 gallons of ordinary water, it *looked* like a pool. In the early 1990s, the county argued that it was a tank, which the lab vehemently opposed because to double-line it would involve a multimillion-dollar renovation, shutting down the reactor, and stopping research for a year or so. In 1993, O'Toole herself had formed a Spent Fuel Working Group to look at issues with spent fuel pools across the DOE complex. Referring to BNL the committee stated, "The [HFBR] fuel canal is unlined and there is no continuous and accurate way to measure leakage." The group was told by the lab that no leakage problems from the pool have ever been detected, including in two wells near the reactor. O'Toole's group, however, did not ask for the location of the monitoring wells that were the basis of this statement; had its members investigated further, they might have realized that the wells were not in the right place. The final report made no mention of the need for more monitoring wells.[44]

From 1982 to 1986 a potable well south and slightly east of the HFBR registered a small amount of tritium, but it was well below the EPA's drinking water standard and thought to be from a leaking sewage line. The well was closed in 1986 after its VOC concentration exceeded the VOC standard and no subsequent investigation was done, partly because the contaminated water would be hazardous waste and partly because two sampling wells, labeled 85-01 and 85-02, were installed in 1989 south and east of that potable well and showed very low levels of tritium. (The local DOE, naturally, participated in all these decisions.) Unfortunately the wells were not directly down-gradient of the HFBR, so not in a place to intercept the new narrow tritium plume.

In 1993, Paquette proposed to install three to five sampling wells south of the HFBR at a cost of $15,000–$30,000. While air monitoring of the HFBR showed no indications of elevated levels, the reactor's groundwater monitoring "could be improved," Miltenberger wrote in a memo, but the existing wells "have not detected any significant tritium concentrations. Consequently, I would still conclude that HFBR specific monitoring wells would be useful but a low priority unless operations data suggest otherwise."[45] In July 1994, Paquette wrote a memo stating that BNL had a responsibility to establish groundwater monitoring where facility operations could impact groundwater. In October, Suffolk County emphasized again that the tank must be abandoned, removed, or relined. Friction intensified between Suffolk County and the lab.

To satisfy the county regulators and allow research to continue undisturbed, Assistant Director for Reactor Safety & Security Sue Davis offered a compromise at a meeting in November 1994: in lieu of renovating the pool, the lab would install two new monitoring wells directly downstream from the reactor. In an extraordinary act of trust—which it would soon bitterly regret—the SCDHS agreed. In a March 1995 meeting the lab said it approved the monitoring well project, but later that spring canceled installation due to a budget reduction. (Here, too, the local DOE office knew of these decisions.)

The process for funding such projects was straightforward. Every year departments and divisions submitted requests for infrastructure projects such as building improvements and worker, fire, and environmental safety measures. A committee evaluated and prioritized the requests using a guideline that the DOE had given the lab in 1992, called the Risk-Based Priority Model (RPM), and the lab funded these priorities as per available resources. The highest priority projects were those with tangible and immediate pay-offs in reducing worker injuries and accidents. To Paquette's frustration, the HFBR monitoring wells were judged not competitive enough.

Finally, in March 1996 the lab approved the wells, using the same plans and estimated costs of two years earlier. Two wells were installed 100 feet down-gradient of the HFBR on July 1 and 2. Their pumps were installed on July 30. These were the wells whose samples eventually alerted Paquette to the leak.

A well, like any instrument, has to be "developed," that is, prepared and calibrated. A new well is usually silty and must be pumped, cleaned, surveyed, and labeled before being used. The lab's environmental staff had mapped the lab as a grid, whose columns were labeled with numbers 1–135 from north to south, and with each box given a number; the two new wells, for instance, were in Box 75. After installation, on August 7, the lab had a licensed surveyor determine the exact location of each well. On August 29, the OER gave each well—and new ones at the gasoline station and motor pool—permanent identification numbers so they could be tracked in the master database; the already existing wells near the HFBR were 85-01 and 85-02, while the new ones were 75-11 and 75-12. Normally, it takes six weeks to develop a well, but the lab's environmental monitoring program was backed up that fall with other air and groundwater samples and samples from the landfill, sewage treatment plant, experimental areas at the accelerator, and other places required by New York State's Department of Environmental Conservation. There seemed no reason to rush sampling of 75-11 and 75-12 as the pool had shown no signs of leakage. Still, Paquette grew frustrated with the delays.

"We therefore 'fit' the HFBR, Gasoline Station and Motor Pool wells into our October sample schedule, which was a bit lighter," Paquette wrote Mona Rowe, in reviewing events leading up to the first week of January 1997. The first samples from 75-11 and 75-12 were taken on October 17, 1996. The analytical lab normally turned around sample analysis in six weeks, but the analysis of these was delayed because of the backlog. "By late November," Paquette continued, "we realized that the HFBR samples were being held up, and we put these samples to the 'front of the line.'" The analysis, which came back on December 5, found that 75-11 had a tritium level of 454 picocuries per liter, and 75-12 a level of 2,520 picocuries per liter, both well below the drinking water standard of 20,000 picocuries per liter, so there was no real concern. Still, the numbers, especially the latter one, were surprisingly high, so on December 11, Paquette ordered additional samples from the two wells—and their analysis expedited—and the results came back on January 8, 1997. These showed a level of 2,110 in 75-11 and 44,700 picocuries per liter in 75-12, the second over twice the drinking water standard. He immediately ordered a resampling for the next morning. These results came back the next day, January 10, showing 6,800 picocuries per liter in 75-11 and 37,600 picocuries per liter in 75-12, the small amounts of increase and decrease probably partly due to the margin of error in the analysis process.

"It isn't pretty," Paquette concluded his letter to Rowe, but "that's the story."[46]

BUCKETS, LASERS, AND FLOATS

The tritium-containing water seemed to be coming from the HFBR building—but where? Reactor Division employees continued to test pipes and drains and found them secure. Two years before, 150 gallons of tritium-containing coolant had spilled in an accident that was reported and investigated, but rechecks showed it had been cleaned up. Lab engineers now focused on the spent fuel pool, and the reliability of the leak test.

Could the water have worked its way naturally through the 3-foot-thick concrete wall or floor, or had the pool been damaged?

On February 10, Reactor Division engineer Ray Karol made a rough calculation of leakage of water through concrete using the original 1963 value of permeability. That showed an expected rate of leakage of five gallons per day through three feet of concrete after eleven years. Then he discovered that that 1963 permeability value used was for air, not water, and when correcting the mistake on February 25 he found that the time before an undamaged three-foot concrete canal would leak would be 115 years—that is, if the pool were indeed leaking, it must have been damaged. Reactor engineers used remote-controlled, zoom-lens underwater cameras to make videotapes of the pool, and saw no visible leakage paths. Later, in December 1997, a more sophisticated calculation, taking into consideration the sensitivity of permeability on the unknown ratio of water to cement in making the concrete changed the penetration time through three feet of concrete to between eleven and thirty-six years for a change of water to cement ratio of about 15 percent.[47] With the variety of unknowns, a calculation is not a reliable guide for determining a leak; there was no substitute for a precise measurement.

Surprisingly, O'Toole's Spent Fuel Working Group committee had not asked how small a leak BNL scientists could measure. They worried about earthquakes, and, like the Brookhaven reactor engineers, had not considered a small leak consequential—socially or health-wise—and thus did not suggest monitoring wells or a liner for the pool.

The spent fuel pool's original design did not include a built-in instrument to measure pool leakage, but Reactor Division operations staff periodically tested for leaks by filling a stainless steel container with water and comparing its evaporation rate with the pool's. It was informally referred to as the "bucket test." "That's a *real* bad term," Reactor Division operations group leader Doug Ports said years later, as it conjures up images of something makeshift and imprecise. But it was a standard test used for such purposes as to indicate a leak, which would be further investigated if initial indications warranted. "It wasn't flimsy," Ports said, just not intended to detect a few gallons of leakage in a 68,000-gallon tank that also evaporated 50–100 gallons per day.

The principle was simple: take a stainless steel flask of the sort you see in chemistry labs, fill it with water, and immerse it in the pool. The pool's water level will go down an expected one to three inches each week from evaporation. Assuming that the temperature and evaporation rate of the water in the "bucket" and the pool are the same, any difference between the two levels means there's a leak. Each time Ports and company had run the test, there was no measurable difference. "We never did an error analysis or an uncertainty analysis, nor did we see the need for a more sensitive test."[48] It was clear, though, that such a test could not see a loss of a few gallons per day, for the change in water level would be tiny, less than 0.05 inches—and such a difference could surely not be detected with rulers.

The operations staff had conducted the stainless steel container tests regularly over several years and found no leakage. After Paquette's discovery, Ray Karol reexamined the test and reported that small changes in temperature could produce differences between the levels in the container and pool. Ports explained, "Suppose the temperature in the pool goes up by a degree. The water level rises; as its density goes down its volume goes up. But there's a lot more water to expand in the pool than in the flask, so the water level rises more—and the opposite for a decrease in temperature. Usually there's not much temperature change in the pool, but we didn't measure it." Another problem was that some of the spent fuel rods in the pool were hot, making its temperature not uniform.[49]

The operations staff sought a more sensitive test. Ports explained, "We needed to control evaporation and reduce it to the lowest level achievable, we needed to know the temperature of the pool and adjust test results based on temperature changes, and we needed to measure small changes in water level. There was a lot of learning at that point." He and other Reactor Division staff covered the pool with a plastic sheet to prevent evaporation, installed a mirror on a float, beamed a laser at it, and measured the position of the reflected spot over time. But uncertainty in the size of the laser spot gave them a bigger error bar than in the bucket test. "We lost a week trying to make it work," Ports says.

Ports and a team of other staff members then installed thermocouples in the covered pool to gauge its temperatures more precisely, put a ruler on a float in the pool about a dozen feet from the edge, and had experts who surveyed accelerator beam lines measure the float's position. The team validated the measurement system by removing and adding a few gallons of water from the pool to simulate a leak and see if they could detect it. They were relieved to find that they could measure a loss down to a few gallons. They ran the new test, measuring every few hours over a week. The first run indicated that the water level in the pool was dropping around a fifth of an inch per week—or 0.03 inches per day, about the thickness of a credit card, much less than that due to evaporation, which varied between 0.2 and 0.4 inches per day, which would have masked the effect of the small leak with the standard "bucket" test. Including the uncertainty in the measured drop in water level, the leak worked out to a rate of between 7 and 14 gallons per day.[50]

Lab officials then asked Ports to conduct a second test to confirm the results of the first. "They really wanted us to get as precise as we could," Ports said. He conferred with the Reactor Division staff for weeks on ways to reduce the uncertainty. In the end the team added more thermocouples to better monitor the pool temperature at different locations, used a heat exchanger to devise stricter temperature controls, and installed a second pool cover about a foot above the first. This more precise test indicated a leakage of 6–9 gallons per day.[51] It was small—but after carrying out what is known as a "source term calculation" Ports and Karol found that such a loss accounted for what was known about the length, dimensions, and concentrations of the plume. The leak rate, combined with data of the distance the front of the plume had moved from the building with a rate of groundwater flow of one foot per day, indicated the pool had been leaking about a dozen years.

Brookhaven scientists had decades of experience and state-of-the-art equipment for measuring minuscule amounts of radiation; they were able to detect, for instance, a few atoms of a radionuclide. But the fact that they had not developed the capacity to detect minute changes in the level of the reactor's spent fuel pool reflected an implicit assessment that an approximate measure sufficed. Given that a leakage of 6–9 gallons of tritium in the groundwater would not get into drinking water or go off-site—and even if it did was not a health hazard—it indeed seemed insignificant. But to elected officials, activists, the media, and many neighbors, failure to develop a test sensitive enough to measure such a loss indicated carelessness and incompetence, or perhaps even a cover-up. The media had a field day after they learned the test's nickname.

A few months later a *Newsday* reporter arrived to interview Ports. Ports outlined the stainless steel container test to give the journalist some background. He believed he had communicated the fact that it was used elsewhere as a standard test for such applications, the extraordinary difficulty in measuring that small a leak given the pool's evaporation, the fact that any leak that could not be detected by such a test would not be a safety hazard, and that as soon as an indication appeared that the test was not sensitive enough he and Karol developed a more precise one. A week later the *Newsday* article appeared. It mocked the test and seriousness of the lab's employees. "Bucket of Trouble," ran the headline.[52]

"That scarred me," Ports recalled years later. At the first good opportunity he left the Reactor Department.

DRIP, DRIP, DRIP

In March, Paquette's complex of wells turned what had been an outline of the extent of the tritium in the groundwater into a detailed, 3-D picture. Starting underneath the reactor building, the plume flowed south with the groundwater, deepening, broadening, and becoming more dilute.[53] Calculations showed that the plume had roughly 15 curies of tritium, or less than that of a self-illuminating exit sign.[54] The numbers indicated to the scientists that the plume was harmless health-wise—but again, as Manowitz had remarked almost half a century before, that did not mean harmless otherwise.[55]

Terry Lash was the lone person who still clung to the idea that the HFBR spent fuel pool was not responsible for the plume, and he prepared a press release saying that the tritium was coming from the lab's sewer system. Lab scientists were dumbstruck. "Check this out," wrote Gary Schroeder to Paquette, "this hypothesis makes no sense." To stop the press release, Schroeder had to drop everything and spend hours with several DOE officials going over maps of the sewage system and the well data to explain why Lash's idea was preposterous.[56]

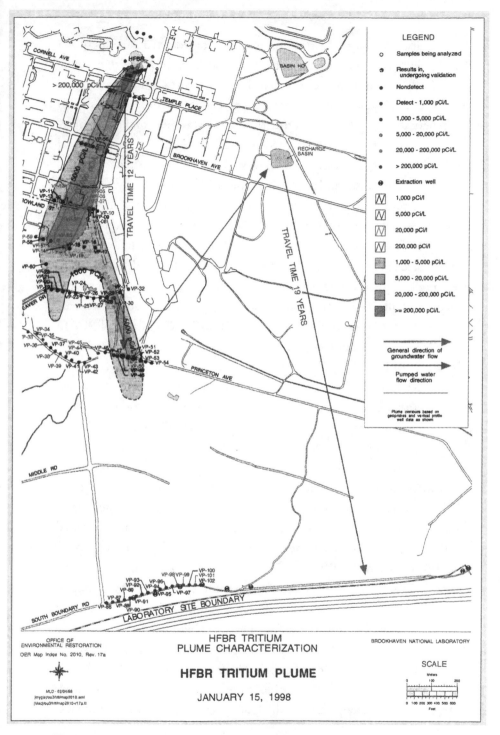

Figure 2.7

Map of tritium plume and concentration, showing remediation path. Water was piped from the leading edge of the plume, sent back upstream, and put back in groundwater.

The discovery of the tritium plume exacerbated a trust gap between the lab, neighbors, and several institutions charged with protecting public safety. BNL officials had declared that the pool wasn't leaking and needn't be double-lined—but now they determined it was leaking, so why should one believe their statements that the tritium leak was small and not dangerous? The discovery also undermined the Suffolk County Department of Health Services, a competent agency, trusted and well-respected by regulators, environmentalists, and public officials, which took seriously its responsibility to protect the county's environment and groundwater. Its officials now felt betrayed, and were infuriated to hear excuses to the effect of, "We didn't really have to live up to the agreement because we are a federal facility and technically exempt from local jurisdiction," or "The wells were expensive, installing them would have stopped research—and besides, what's the harm?" The Suffolk County Water Authority desperately ran advertisements in *Newsday* and other local newspapers: "PUBLIC WATER SUPPLY IS SAFE: No Contamination From Brookhaven National Laboratory in Public Water Supply."[57]

The plume had also embarrassed the DOE, the federal agency that financed and supported the lab. Its management and practices had been severely criticized by the DOE's own report—"Task Force on Alternative Futures for the National Laboratory," known as the Galvin Report after its head—just two years previously. Furthermore, much of the DOE's effectiveness depended on its reputation, for the agency had to work with governors, congressmen, and various industries to whom it had to make reliable promises about its facilities—their costs and deadlines, and the associated public and worker safety and health. Finally, the DOE wanted to protect its investments. The HFBR was a valuable and costly scientific instrument, especially in view of the increasing shortage of neutron-scattering facilities. While the HFBR had cost $12.5 million to build, it would cost around $3 billion to replace. For this reason the DOE had been seriously considering the lab's request to upgrade the precious instrument. In mid-January, Patricia Dehmer, the director of the Office of Basic Energy Sciences, which funded and evaluated the HFBR's scientific programs, was sitting in her office with several Brookhaven officials discussing how to go about the upgrade.

Figure 2.8

Suffolk County Water Authority notice, placed in local newspapers to counter charges that it had ignored Brookhaven National Laboratory's groundwater testing. Courtesy Suffolk County Water Authority.

"Someone came to the little conference room and told us of the tritium leak," Dehmer rememberd. "That was the end of the proposed upgrade."[58]

Rebuilding trust was not easy. O'Toole and Lash's first attempt was to instruct the lab to be transparent and to openly and regularly communicate all its findings after DOE approval, aiming to show that the lab was hiding nothing. Initially, the effect was the opposite. Issuing frequent press releases about well results and efforts to find old chemicals in the ground made it hard to distinguish significant from insignificant, and new from old, findings. The transparency reversed the normal scientific investigative process in which each data point is treated as a puzzle piece. Methodically fitting those pieces together reveals a picture, and then the news event is to announce that picture—the nature, size, and hazard of the plume in this case. However, that process could not happen now. In this case, each data point became a separate news event, making the lab's investigation appear haphazard, the lab's knowledge of the plume constantly changing, and its scientists floundering and untrustworthy. Politicians and reporters could assign each data point to whatever picture they wanted.

Paquette's plan for drilling wells compounded the problem. It was efficient, cost-effective, and scientifically sensible. He first drilled wells near the reactor to find where the plume started, then worked downstream. That way, he used knowledge of the groundwater flow to guide installation of the next set of wells. This collected information about the plume in the quickest and most direct way. But in the new transparency, the findings of each new set of wells were released as they came in, and headlines screamed that the plume was growing in large amounts daily as Paquette zeroed in on the hottest spots—eleven and then thirty-two times the drinking water standard. The tritium seemed to grow increasingly more dangerous, and because of the way wells were drilled it looked like it was growing worse and marching much faster than the foot-a-day groundwater flow, toward the southern communities.

It made for exciting journalism. Finding dirt about an industrial plant can require digging and the need to fight stonewalling by an institution trying to hide its guilt—but not for a federally supported civilian laboratory for whom it was not an option to refuse. Some reporters called daily, even several times per day, knowing there would soon be a new batch of data, another story about a new leading edge or concentration. "Tritium Levels Rise," ran a typical *Newsday* headline; "Lab's Toxic Plume Wider than Thought" another.[59]

The most dramatic single instance of what might be called information shakedown occurred on March 4, when O'Toole and Lash came to Brookhaven on a one-day visit, intending to try to bolster employees' plummeting morale. The event started smoothly. Lash thanked the employees for their hard work. O'Toole said that BNL had gone through a long period of problems and was now moving on. She said that there would continue to be plenty of regulations and scrutiny going forward, but said she was supportive. She asked employees to be patient and not get "huffy and defensive" in reacting to oversight.[60]

That morning, however, a new set of data happened to come in. Normally, scientists would wait for a full set of data from other wells, and integrate that with the existing information to improve their picture of the plume—which was, after all, their goal. The news event would be about the overall story of the plume. That became impossible as, by noon, a reporter managed to get his hands on the fresh data in isolation from the rest and asked the lab for confirmation. The lab's media relations officials could not deny the story, but also did not want the reporter to have an exclusive, which would enrage other reporters, so they rushed a press conference for 4:00 p.m. that afternoon in the Berkner Hall lobby and invited the other reporters covering the plume so they could all have it at once. While O'Toole delivered the information, media relations staff looked around to see which reporters were absent, then frantically dialed them to relay the news and faxed a hasty, six-sentence press release.[61] The announcement was not about the plume; it was about the numbers. The hurried way the press conference was thrown together made the lab look unprepared—as though it had just received unexpected and shocking news. That day, a local radio station, WRCH, did the first of what became a series of parodies of "Crookhaven Lab."

Media coverage was shaping public perception. For those who knew about the lab only from headlines, the lab's work seemed not science but endangering health and safety. When an outraged scientist called the office of Michael

Caracciolo, the area's representative in the Suffolk County legislature, to ask if the lab really posed a health and safety problem, as the legislator was claiming, Caracciolo's staffer replied, "Yes! Don't you read the newspapers?"[62]

The lab issued an explanatory bulletin, "Why Do We Continue to Have 'Revelations' about Contamination and Other Problems at BNL?", but that did not stop headlines.[63] Full transparency created other problems. Politicians and reporters often saw differing measurements of extremely low levels of tritium as indications of cover-ups or incompetence rather than the difficulty of analyzing samples at the limits of the instruments' sensitivities.[64]

Other revelations came from the lab's Historical Legacy survey as part of the Superfund process search for old sites of contamination. New sources were discovered and known ones revisited—and press releases had to be issued for each, even if the contamination was insignificant, making it seem like the pollution was ongoing and unstoppable. A few days after O'Toole and Lash's visit, test results from sampling the contents of an underground tank built in the early days of the lab showed some small contamination of strontium-90 and tritium, and a press release was issued. Pumping of the tank was completed two weeks later, accompanied by another press release. Updates were regularly issued for the ongoing investigations, sometimes more than one per day.[65] On March 10 a lengthy press release summarized the information about the tritium plume obtained so far, with updates on March 14 and 21. Two press releases were issued on March 27, one an update on the plume, the other on the underground tank. "Drip . . . Drip . . . Drip . . ." ran the cover story of a section of *Newsday* the following week.[66] Neighbors might well wonder if their area would become another Love Canal, Woburn, Toms River, or Hinkley.

The lab's troubles attracted the attention of Robert L. Park, a physicist who wrote *What's New*, a one-page weekly newsletter. It was widely read as the conscience of the physics community, for it said things that everyone knew but were afraid to say for political reasons. Aside from covering scientific developments, *What's New* pilloried alternative medicine and pseudoscience—and Park loved targeting congressmen who fell for them, especially those on appropriations committees. While the American Physical Society officially sponsored *What's New*, Park made the APS so nervous that it forced him to add a disclaimer. At the bottom of each newsletter Park therefore put the following: "Opinions are the author's and are not necessarily shared by the APS, but they should be."

"Why must residents find out what's going on in dribbles?" Park wrote in a *Newsday* editorial. In *What's New* he predicted, "The steady drip of new disclosures is rocket fuel for political demagoguery."[67]

"WE'RE IN TROUBLE"

By this time, many scientists at the lab feared for their research. Joanna Fowler, the head of Brookhaven's positron emission tomography (PET) scan program and one of the country's leading PET scan researchers, was determined to keep her program from getting the same media and political treatment as the HFBR. An inveterate organizer and cat fancier, she and others had stopped the lab's plan to round up and kill on-site feral cats by coordinating members of the chemistry, medical, and other departments in a mammoth catch-and-neuter effort. In March she organized her Center for Imaging and Neurosciences in a cleanup effort. Her mentor Al Wolf was in a nursing home, and the department had inherited his huge collection of chemicals, some toxic and others radioactive, which he had meticulously logged in a beautiful Moroccan leather notebook. Wolf had worked at Los Alamos on the Manhattan Project to build the first atomic bomb during World War II, and his souvenirs, including "bomb glass," pieces of sand turned to glass by the test explosion, were still slightly but harmlessly radioactive. Among Wolf's other work, he had pioneered and patented a technique to use tritium to label compounds such as proteins and peptides used in research involving reproduction, cancer, and shellfish toxins.[68] Others in the Chemistry Building had flasks with solutions containing uranium or other radioactive substances. If tritium could make people talk about closing the lab, Fowler feared, so could many of the Chemistry Department's possessions and routine analytical procedures. "I worried that if some of this got out or on somebody's shoes or in their computers, and the newspapers got wind of some tiny amounts of radionuclides, it would spell the end of our research," she recalled. "Even a harmless leak could be disastrous! A single atom could do us in. I didn't want to give them that bone."

Fowler created a "Looking for Trouble" group, a team of investigators who, once a week, scoured Chemistry Department rooms and labs to ask their occupants, "What's the worst thing that could happen in what you are doing?" Nothing was overlooked. "Chemistry labs have a lot of thermometers," Steve Dewey recalled, "and you often kept yours in a beaker next to the sink. All you had to do was knock one over, it would break, the mercury [a toxic material] would go down the drain, and that would be it." Fowler's group set in motion a massive department cleanup. Most of her colleagues were cooperative, but many were annoyed. "It did make things cleaner, and more secure and accident proof," Fowler said.

On March 21, Lash and O'Toole returned to the lab, accompanied by Krebs, with two pieces of encouraging news. One was that the DOE had a plan to ship the spent fuel rods in the HFBR's pool off-site by the end of the year so the pool could be relined for the restart. Krebs told Reactor Division employees, "Right up front, I want to restart the reactor because it is clear to me that it is the right thing to do scientifically, for the Office of Energy Research, and for the neutron-science community in this country—but I cannot guarantee that it can be done."[69] O'Toole reassured the employees: "I can't imagine any plausible or even remotely possible way that this tritium leak can be causing any kind of health effect on or off site. If you don't have exposure, you can't have health effects, and we don't have exposure. There is no data that I know of—and I have examined data from the 1980s through today—that indicates that there would be any reason to expect that there are any public health consequences from Brookhaven releases or operations."

During the March 21 meeting, lab official Robert Bari glanced at the three DOE officials behind the same table. He knew that O'Toole had a reputation for severely punishing DOE labs for environmental infractions, that Lash had been outraged by AUI's actions in the Lisbon initiative, and had heard that Krebs felt snubbed by lab leaders. "We're in trouble," Bari remembers thinking.[70]

The second piece of good news was that the DOE had approved the "Valentine's Day Massacre" plan. In the "Tritium Remediation Project," the lab would siphon out water whose levels of tritium were already below EPA drinking water standards—drinkable water—filter it, and send it up-gradient into a recharge basin used for environmental remediation projects. In heading back south, decay and dilution would reduce the tritium still further. The plan seemed ridiculous to Representative F. James Sensenbrenner (R-Wisconsin), the incoming chairman of the House Science Committee. Sensenbrenner was appalled at both the remediation project and water hookups, saying that the DOE had identified neither the need nor the funds for them, and that it was a lot of taxpayer money to spend pumping drinkable water up from the ground and putting it right back down again.[71] He came to the lab in June and discovered that the hookups were for chemical plumes, not tritium; still, he was annoyed about the money and the calls for the lab's closure. Someone anonymously wrote the US Inspector General's office to leak the news that the federal government was wasting taxpayers' money on the tritium project, but DOE lawyers fended off the charge. Still, the tritium project made O'Toole and Lash "much more optimistic and confident in our ability to arrest the tritium leak and resolve the problem, and in BNL's response to the problem, than a week ago."[72] A week later the Suffolk County Department of Health Services (SCDHS) approved it in principle.[73]

The lab's Plant Engineering workers began installing three extraction wells 3,500 feet south of the HFBR to siphon water from the leading edge of the plume. The system was to "draw a line in the sand," literally; while it initially pumped essentially clean water 3,000 feet back upstream to the recharge basin, it guaranteed that if somehow the plume traveled to that point it would not go further. From the recharge basin, it would reenter the ground and virtually disappear by decay and dilution in the nineteen years it would take the groundwater to flow to the boundary. Plant Engineering completed the DOE-approved system on April 17, three days ahead of O'Toole's sixty-day milestone deadline. The next day, the OER submitted a report on the technical details of the system to Suffolk County authorities. But before pumping began O'Toole abruptly put the action on hold; the Suffolk County authorities told her that state and EPA regulations forbade putting contaminated water back into the ground. The county also wanted to know more about the plume. Startup of the system would have to await a variance.

In March, each of the three institutions struggling to control events—the DOE, BNL, and AUI—was in the process of changing leaders and in some managerial disarray. A new secretary of the Department of Energy, Federico Peña, was sworn in on March 12; Brookhaven's Director Nicholas Samios announced on March 7 that he was stepping down from the lab effective April 30; and Lyle Schwartz took over as the new AUI President on March 17.

* * *

Peña became the first Hispanic mayor of Denver in 1983. During his term the FBI raided—"invaded," workers called it—the Rocky Flats weapons production plant about twenty miles northwest of Denver, and he was highly aware of how serious environmental contamination can be at federal facilities and the dangers of not responding forcefully. In 1993 Peña became President Bill Clinton's secretary of the US Department of Transportation, during which time his reputation as a leader was sullied for failing to take action against a company with a bad safety record. On May 11, 1996, Valujet Flight 592 had crashed near Miami. After arriving at the scene of the crash, near the bodies of the 110 victims, Peña had declared ValuJet to be a safe airline. Documents then surfaced revealing that ValuJet's safety practices were so poor that the Federal Aviation Administration had wanted to ground the plane. This was the first major safety incident that he encountered as Energy Secretary, and he was understandably edgy. He had little experience with energy issues, and in 1997 was taking charge of an agency known for infighting.

* * *

Nicholas Samios was born and raised in Manhattan, the child of two Greek immigrants. He studied physics at Columbia University (BA 1953, PhD 1957), and joined Brookhaven's Physics Department in 1959. He led research groups that made important contributions to high-energy physics, and became chairman of the Physics Department in 1975. AUI had chosen him to become laboratory director in 1982 during a time of turmoil when its major new accelerator project was about to be terminated, and he had miraculously converted the partially constructed project into a new major facility, the Relativistic Heavy Ion Collider (RHIC), with a different funding source. Samios privately informed members of the AUI board that he intended to step down after he turned 65—which would be March 15, 1997—and by the anniversary of his fifteenth year in the position, which would be the end of April. He had expressed this desire to the trustees in January, but they had asked him to hold off, because they were in the process of looking to replace their own long-term president, Robert Hughes, and they did not want to conduct two high-level searches at the same time. Now that Lyle Schwartz was poised to take over, Samios prepared to officially resign as director.

But Samios's resignation became absurdly tangled in politics, journalism, and institutional narcissism. He arrived in Washington late on March 6 with a letter to give AUI trustees the next morning announcing that he was stepping down as lab director; he would simultaneously email it to lab employees. That evening AUI officials notified DOE and congressional officials, who in turn informed D'Amato and Forbes, thinking to appease them. But the congressmen wanted credit for Samios's departure and leaked it to *Newsday*.[74] Meanwhile, the DOE thought its pressure had caused Samios's resignation. Hughes was annoyed because he felt he should be the first to inform lab employees, and issued his own announcement.[75] At Brookhaven, even simple actions were no longer simple.

Beforehand, Samios had taken two forward-looking steps. One was to appoint a committee to review the decision-making process for the lab's ES&H activities.[76] Headed by Robert Bari, the five-member committee included a representative of Suffolk County's Department of Health Services. The other was that he initiated a reorganization of the lab's Public Affairs office, leading to the hiring of Marge Lynch as manager of that office in mid-April, reporting directly to the lab director. An experienced public affairs expert, Lynch had worked for Northville Industries in 1988 when the company discovered that around a million gallons of leaded gasoline had leaked into the ground from a hole in an underground pipe at its storage terminal. Lynch had third-party professionals run the public meetings to ensure that the tone of the meetings was serious and steady. Lynch set out to consolidate and refocus all the lab's communications functions; she made Geiger manager of community relations, and Rowe became manager of media and communications. Lynch handled government affairs herself, working with the lab directors. Soon after stepping in, Lynch wrote:

> The lab's credibility is on the line with its neighbors, civic and political organizations, public officials, and other key stakeholders. . . . The most important part of any initial strategy that BHG and BNL creates is consistency of message, coordination of effort, and careful prioritization of contacts/briefings with civic leaders and individuals concerned about the lab. As an example, Bill Smith despite his lung capacity is not an "affected" member of the community, and should not receive star status at the expense of other stakeholders. BNL is in the process of integrating all of its communications functions.[77]

Two days after Samios's announcement, *Newsday* ran an even-handed article about his legacy—"Dueling Legacies," as the headline put it. Samios had a good "taste" for the right problems in experimental physics and an "unwavering devotion to first-rank science," the article said, but "never developed the same feel for handling public concerns over environmental problems." The article quoted a civic association leader as saying that they had "good relations" with the lab under Samios, and also quoted leading Brookhaven scientists on Samios's role in building up the lab's biology program, addiction research, high-energy physics, and other areas. "I think he's one of the truly great experimentalists of the 20th century," said Samuel C. C. Ting, whose Nobel Prize–winning discovery took place at Brookhaven. "He should have gotten the Nobel Prize a long time ago." But the article also quoted Robert L. Park as saying that Samios "wouldn't be the first administrator who underestimated the fear factor in the public."[78]

* * *

The new AUI president, Lyle Schwartz, had studied science engineering as an undergraduate at Northwestern University (1959) and materials science as a graduate student there (1964) at the beginning of exciting years when the Soviet Union's launch of Sputnik suddenly unlocked US funds for materials science. He took up power weightlifting as a graduate student to relax from his X-ray diffraction research, and became a national champion. He was director of the Materials Science and Engineering Laboratory at the National Institute of Standards and Technology (NIST) from 1984 to 1997. He liked working at NIST, which he found to have an excellent management structure for environmental health and safety.

In February and early March, as the lab's tritium problem worsened relations between AUI, BNL, and DOE, Schwartz wondered about the wisdom of accepting the new position, but he arrived determined to bring to the lab the same approach to environmental management he found at NIST, and with Leland Willis, AUI's environmental lawyer, and others, worked out a detailed, three-level Management Systems Improvement Plan (MSIP).

Schwartz's first responsibilities included trying to keep AUI and DOE from fighting over Samios's successor. Panicked about its public perception, and under heavy pressure from D'Amato, the DOE was leaning on AUI to choose someone outside the lab. AUI, already furious with DOE's power grabs over the lab, was equally determined to show that it was in charge. AUI's Search Committee, chaired by Cooperman, approached some eminent external candidates,[79] including John H. Marburger III, former Stony Brook University president and former chairman of Fermilab's manager, the Universities Research Association; and D. Allan Bromley, dean of the Yale Faculty of Engineering and former US Presidential Science Advisor. Both declined. At the end of March, with Samios's departure looming, AUI realized it would have to find an Interim Director.

WHOSE LAB IS IT?

On April 1, Schwartz called Physics Department Chair Peter Bond from the AUI's Washington headquarters. With Samios leaving at the end of the month, Schwartz said, AUI needed to find an interim director for the lab. The normal practice would be to elevate a senior administrator, but Schwartz told Bond that both the DOE and Senator D'Amato were dead-set against it, but also that AUI's trustees were adamant that the decision was theirs. Schwartz and Bond batted about ideas.[80]

A week later, Schwartz and Bond had dinner at Lombardi's, an Italian restaurant not far from the lab. The matter was now urgent, with the trustees' meeting only a week away. The trustees had authorized Schwartz to defy D'Amato and the DOE and offer the position to Bond.

Bond had been chair of the Physics Department for a decade. He had graduated from Harvard in 1962 and received his PhD from Case Western Reserve University in 1969 and, following a postdoc at Stanford, joined the lab in 1972. Prior to becoming Physics Department chair in 1987, Bond had worked in heavy ion physics and had been part of the first relativistic heavy ion physics experiment at the AGS in preparation for RHIC. The previous three lab directors—Samios, George Vineyard, and Maurice Goldhaber—had each been chairs of the Physics Department. Bond was a natural first choice as interim director.

Bond agreed to have his name put forward, but pointed out that the DOE would have to agree. Officially, that meant Secretary Peña's blessing, but in practice it meant Krebs's, and even more importantly O'Toole's, whose voice as head of Environment Safety and Health was now strongly influential in the DOE, and who had driven the events thus far. Schwartz arranged for Krebs and O'Toole to interview Bond the next day, April 10.

O'Toole happened to be at the lab for a DOE-mandated "Technical Review Meeting" on April 9–10, an expert panel of dozens of people charged with reviewing "the appropriateness of the proposed technical path" for the tritium plume, attended by representatives of BNL, DOE, US EPA, NYS DEC, SCDHS, USGS (United States Geological Survey), the NYS Attorney General Dennis Vacco's office, the Town of Brookhaven, the offices of D'Amato and Forbes, the State Oversight Committee, AUI, and others.[81] Its purpose was to analyze all data from all wells, study models of the plume, review the technical details of options for handling it—and then get the representatives of all those diverse organizations to agree on what to do.

The first day was frustrating and tense, for some attendees had no scientific background while others were experts at remediating toxic substances and had to bring the others up to speed on such basic concepts as half-life. "Certain people had a hard time grasping that the tritium would go away by itself," Gunther remembered.[82] The second day was also difficult, as some experts were flabbergasted that the DOE was seriously considering spending significant amounts of money on a non-threat, especially when Hanford, Savannah River, and other DOE-managed places had plumes orders of magnitude bigger and more toxic than Brookhaven's. County officials had to strongly pressure holdouts; politics, not science, they said, was driving the decision, and the very existence of the lab was at stake.

Bond met O'Toole during a break on the second day, and spoke by phone with Krebs. Both interviews seemed to him to go well, and the lab's media relations department drafted press releases announcing that Bond was the lab's new Interim Director. Bond traveled to MIT the next day as a member of a visiting committee. There he ran into Robert Birgeneau, the dean of science at MIT and an AUI trustee, who congratulated Bond on his proposed position. Bond returned to Long Island over the weekend shaking his head at the speed with which the news had spread.

Bond wasn't in consideration for long. On Monday, at a hospital with his wife, who was recovering from an operation, he called Schwartz—who told him that, hours before, O'Toole had vetoed Bond. The decision was not about Bond personally, O'Toole later told Bond; it was that Bond was "part of the AUI package and the strong belief was we needed much more coherent and enlightened leadership than AUI could provide."[83]

The next day, with only two days to go before AUI had to decide on an interim director, Schwartz tried desperately to come up with a plan that would work for both the DOE and the trustees. Boxed in a corner, he proposed that, if the DOE continued to reject Bond, he himself could step in as director to emphasize AUI's commitment to fixing the lab (plus as new to AUI he was effectively an outsider, as the DOE desired), and Bond and Michael Bebon would become deputies for science and operations, respectively. Schwartz met Bebon for dinner and called Bond. Both agreed.

Schwartz relayed the week's developments to the AUI trustees two days later. Krebs also came to the trustees' meeting—O'Toole was supposed to come but the trustees were told she was ill—to explain what the DOE wanted and why. Krebs told the trustees that the scientific community had to recognize that, as the AUI minutes report it, "things are not so simple or as straightforward as they were thirty years ago under the AEC" and "are likely never to be as simple again." She said that "the Laboratory's story is not told as well as it could be, both within the DOE as well as in the outside community," and intimated that the DOE secretary—who according to the contract must ultimately ratify the contractor's choice for laboratory director—might ask to interview all of AUI's candidates.[84] The trustees were furious and talked back. "Voices were raised," said the normally even-tempered Krebs. "I raised *my* voice."[85]

The trustees did not take it well. Whose lab was it, AUI's or the DOE's? Some said that "AUI must demonstrate that it, AUI, is in charge and that it has been effective in helping to put in place management changes that will ensure integration of ES&H into all Laboratory activities." Others noted that AUI's demonstration of its leadership would be especially important if the DOE insisted on recompeting the contract when it expired in 1999.

What enraged the AUI trustees, though, was the DOE's veto of their choice of Bond. They considered defying the DOE by proposing Bond anyway, with the fallback of having Schwartz as interim director. Some trustees even objected to a fallback, saying that it would "allow DOE to control AUI's decisions." After heated debate, the trustees backed away from defying the DOE and voted to propose Schwartz as Interim Director, with Bond and Bebon as Interim Deputy Directors, with the backup plan having Bond as the interim director. The trustees forwarded this to the DOE for approval on April 18.[86]

After Krebs left the board meeting, she met with the lab's upper management but without Samios, and told them she supported the lab but not its current leadership. She complained that when Samios came to Washington he never visited her or other DOE officials. Burton Richter—the director of the Stanford Linear Accelerator Center—"does it right," she said; when he comes to Washington he "schmoozes," that being a time-tested way of forging trusting relationships.[87]

In principle, Schwartz would have been an outstanding choice as lab director. He was an eminent scientist and administrator, new to the scene without strong historical ties to either AUI or BNL. Thanks to his ideas about environmental management gained at NIST he represented exactly the kind of change that the DOE wanted, and he had already worked out a detailed plan to revamp the lab's environmental management. But Schwartz was unenthusiastic about his dual role. He did not look forward to commuting between Washington and Long Island, and he was aware of potential conflicts of interest in his role as the head of two different institutions. But the DOE's rejection left him hanging. He consoled himself with the thought that the word "interim" meant he wouldn't be long in that position.

Meanwhile, Schwartz crafted plans to revamp the lab's environmental practices to make them fit what the DOE said it wanted – working with Bond and Bebon on his Management Systems Improvement Plan and setting up consultations with the DOE's experts. On April 22, he invited Bill Shipp and Walt Laity of Pacific Northwest National Laboratory (PNNL) to discuss their experiences in 1993–1994, when the DOE threatened to take over the lab unless it made substantial changes. PNNL had fumbled its response until it imported the DOE's experts and then followed their advice.[88] By copying the DOE's processes, Schwartz assumed he was executing the responsibility that the DOE had indicated it was investing in him.[89]

Schwartz also began calling to Peña's office for an appointment. All went unanswered, and the DOE continued to ghost him for the next ten days. He was alarmed. For good reason: unbeknownst to him, the DOE's thinking was turning radical.

NO GET-WELL PACKAGE

In his Senate confirmation hearings, Peña had emphasized that the DOE's facilities must listen to the concerns of their neighbors, and live up to its promises to them. After his confirmation the first trip he made as secretary, on April 17, was to Los Alamos National Laboratory, and beforehand he met with leaders of the LANL surrounding community. He therefore paid close attention when O'Toole told him of Brookhaven's broken promise to the county to install monitoring wells.

For the moment, the DOE planned to slap AUI's wrists at a press conference at Brookhaven on April 29, at which they would discuss the two ES&H reviews: the DOE's Integrated Safety Management Evaluation (ISME) report and the lab's Bari report. At the conference, both the DOE and lab officials would "accept responsibility" for the faults discovered by the two reports, publicly flog themselves, and commit themselves to self-improvement—a common ritual for beleaguered government agencies. O'Toole and Krebs would represent the DOE; Secretary Peña himself would not need to attend.[90]

O'Toole intervened. She had been the target of D'Amato's Godfathering enough to know that anything with AUI in it would be unacceptable, and began lobbying Peña to cancel AUI's contract outright. "I said, 'Look, I think what you should do is put a new person in charge of Brookhaven, and give them a clean sheet and start over.'" O'Toole wanted the DOE to move fast, knowing there would be pushback.[91] She drew up a ten-page document outlining various options for dealing with AUI, "Management Problems at Brookhaven National Laboratory," with input from Lash, Krebs, and two other DOE officials.[92]

One option was to leave AUI in place but threaten to force it to recompete the contract when it expired in 1999, a second was to terminate AUI outright and choose another contractor to take over as soon as possible, and a third was to terminate AUI's contract pending a recompetition. The DOE could fire AUI for two reasons. One was "cause," which would require the agency to show that AUI had breached the contract, and the other was "convenience," or lack of satisfaction at AUI's performance. The document ruled out the former, which would require the DOE to give AUI the chance to correct its alleged failures—and invite a lawsuit in view of the high marks the DOE had given AUI for management.[93]

The options document listed several arguments in favor of termination that had less to do with Brookhaven than with the DOE's own reputation and management ambitions. Terminating AUI would "send a strong message within the DOE complex, particularly at the laboratories, which could have the positive effect of demonstrating DOE's seriousness about holding contractors to strong environment, safety and health performance standards."[94] Another argument was that failure to terminate AUI "may be viewed negatively by the community, local officials, Senator D'Amato and Congressman Forbes." Arguments against termination included the remarkably prescient observation that this step would cause "programmatic disruption" at the lab, another that forcing a recompetition would require the DOE to devote much "thought and sustained engagement" to the process—an undesirable corollary.

On April 24, Peña met with his staff to discuss the options document. O'Toole was proactive, and argued strongly in favor of firing AUI. Lash was ingratiating, assuring Peña that if he decided to fire AUI the recompetition process could be sped up without disrupting either the DOE or the lab. Krebs was circumspect, reluctant to fire AUI. Peña told everyone that he was going to think it through overnight.

In the DOE's eyes, AUI—sometimes mockingly referred to as "six men in a phone booth"—was the weakest of all its contractors. It was nominally an association of nine blue-chip universities that represented a large part of the American scientific establishment. But AUI was now effectively run by those whom the universities designated, rather than by the universities themselves, and so it lacked their institutional backing and resources. The universities thus had not intervened to address the lab's current problems, leaving AUI to cope with them alone. Thus AUI was a far easier target than other more committed university lab contractors such as the University of California or the University of Chicago, or industries such as Lockheed-Martin.

AUI was an easier target for a second important reason. When a contractor gets in trouble after things like an accident or scandal, word gets around and competitors prepare to challenge them in rebids. The DOE starts receiving irate protest letters and phone calls from the contractor's congressional delegation. The contractor then typically puts together what's called a "Get Well" package that outlines measures to rectify the problem, usually including a change in senior management. But nobody would be calling on behalf of AUI or the lab, least of all the lab's political representatives.

The April 24 meeting was in the DOE's headquarters in the Forrestal building. Afterward, Peña's Chief of Staff Elgie Holstein recalled, everyone got up and left the conference room except he and Peña.

> I would normally not remember exact words in a conversation that long ago. But this time I do. We had a panoramic view overlooking the old Smithsonian castle, with the Capitol building in the distance. There was a fire far across town somewhere, with black smoke rising. I was not going to pressure him, but

I did ask him how he was feeling about this, and whether there was anything else he needed, which is part of my job. Without looking at me—he was still looking out the window—and without saying definitively what he was going to do except that he was going to take a night to think about it, he said, "Elgie, after all, what are we here for?" I didn't know him well at that point, but I knew what he was thinking. He was clearly troubled and feeling the weight of that decision. But I had no doubt when I went home that night what he meant and what he was going to do. Sure enough, when I came in the next morning the ball was rolling.[95]

Peña and the others at the meeting, though, had not fully thought through some of the consequences. Most immediately, who would be Brookhaven's director? O'Toole had vetoed Bond. Schwartz, the only other prospect in sight, was president of the organization that the DOE was about to fire, and Peña hadn't bothered to return his calls. The DOE was in the farcical position of either having to leave the lab director-less on May 1 or approving Schwartz on the same day it was firing his organization. The DOE decided that the second alternative was the least ridiculous. Schwartz, who displayed exactly the kind of environmental awareness that the DOE said it wanted as head of the lab, would have to be the collateral damage of a bigger drama.

On April 26, Krebs finally called Schwartz—at home—to tell him that the secretary had approved him and his deputies. She did not say that the secretary was also about to fire him.

Two days later, on April 28, BNL and AUI—still clueless—issued a joint press release cheerfully announcing that as of May 1 Schwartz would take over as the lab's new Interim Director, with Bond and Bebon as Interim Deputy Directors. The three would work as an "integrated team to manage the Laboratory with a new emphasis on environmental, safety and health programs."[96] At the press conference, Schwartz was sober, articulate, and clear-headed, but not without optimism. Inspired by NIST's outstanding handling of environmental health and safety, he promised a "culture change that will raise our environment, safety and health programs to the same level of excellence as our scientific programs." He closed by referring to two ancient Chinese sayings: the one about the "good fortune to live in interesting times," which many people had already cited to him, and the other, which he found more

Figure 2.9
Incoming laboratory managers approved by the DOE. Left to right: Peter Bond, Lyle Schwartz, and Michael Bebon. On May 1, 1997, the agency both appointed them and told them that their positions were effectively terminated. From Newsday. © 1997 Newsday. All rights reserved. Used under license.

pertinent, that "A journey of a thousand miles begins with a single step."[97] He did not yet know that he wouldn't even be allowed that first step.

Senator D'Amato went ballistic. O'Toole recalled, "I got an 8 a.m. call from D'Amato one morning [April 29], full with f-bombs, saying, 'You know what those bastards have done? They've put in a new director. Why the hell are they doing this without talking to me?'"[98] D'Amato told reporters, "It's not sound. It doesn't make sense. . . . I think it's shocking. I'm awed. They have found a way to continue to lose public confidence . . . what you need is someone who is not beholden to the AUI." He added, "I am certainly going to tell Dr. O'Toole and the Department of Energy people that it's absolutely imperative that the permanent person . . . be someone from outside." He added, "I'm glad it is interim, let's put it that way."[99] D'Amato did not realize the DOE's predicament that had all but forced the agency to make the choice—nor did O'Toole enlighten him.

That day and the next, two reports on the lab's handling of environmental health and safety appeared, one written by the DOE, the other by Brookhaven. One was O'Toole's final Integrated Safety Management Evaluation (ISME). The report said that the lab was implementing promising new measures to address health and safety, and that nothing warranted "curtailment of Brookhaven National Laboratory operations." The report, though, pointed to weaknesses in the DOE headquarters, its Chicago operations office, and its area office. It identified seven "opportunities for improvement" at both the DOE and BNL. Still, it found that "Although remaining weaknesses need to be resolved, the current Department of Energy and Brookhaven National Laboratory actions to eliminate the source and remediate the tritium contamination have been aggressive and appropriate."[100] The report gave no premonition, or reason why, the ax was about to fall on AUI alone.

On April 29, the Bari report appeared, the lab's own evaluation of its environmental safety decision-making.[101] Parts were damning for the DOE, especially a revelation of the DOE's culpability in the events that delayed the finding of the tritium plume. When the lab had given low priority to the HFBR monitoring wells, the report pointed out, it was following the DOE's own recommended procedures, the Risk-Based Priority Model (RPM), for prioritizing ES&H projects. The lab's failure to fund the HFBR monitoring wells in 1994 and 1995—failures that O'Toole had cited in her testimony to the Suffolk County Energy Committee in February as examples of "inappropriately low levels of the lab leadership"—were due to the project's "very low" RPM scores. "[A]s the hard lesson of the wells has demonstrated," the Bari report said, "one cannot base decisions on this process alone." It provided 16 recommendations for improving ES&H decision-making and underscored the need for the lab to be proactive and to foster a "positive ESH culture."

In short, the lab had erred by following the DOE's own procedures, and should have taken a more enlightened approach to environmental safety than the agency itself.

Together, the ISME and Bari reports provided the framework from which, in ordinary times, AUI would have fashioned a "Get Well Package." But AUI was still in the dark.

Meanwhile, over those same few days, the DOE was rolling out its plans to fire AUI. The secretary's office called the DOE's area office to cancel the slap-on-the-wrist event on April 29 and rescheduled it for May 1, featuring Peña himself, but did not explain why.

Schwartz knew nothing of this, and when he heard that Peña was coming to the lab for a press conference—Peña's first visit to BNL since taking office—he knew some dramatic announcement was afoot. Schwartz assumed it would be a version of the PNNL episode—that the DOE would threaten to recompete AUI's contract—and he therefore brushed up on it. With the plume of no health threat, and given AUI's excellent performance rating, he did not think for a minute that AUI's contract was at risk. "There wasn't anybody at AUI who anticipated DOE's decision to fire us," Schwartz recalled. "So AUI didn't even think about the possibility of a serious lobbying effort against that decision."[102]

Late on April 29, Schwartz and AUI chair Paul Martin received calls instructing them to fly to Washington to meet with Peña the next morning. A reasonable assumption would be that Peña, one new appointee, simply wanted to meet Schwartz, also a new appointee, before the press conference. Schwartz brought along his Management Systems Improvement Plan. AUI had prepared a sunny press release for the next morning announcing the lab's new leadership team, and that in his MSIP the incoming interim director Schwartz had initiated an "attic-to-basement" review of the lab's facilities and procedures, to include participation of the DOE, the New York State DEC, the SCDHS, and the EPA.[103]

But when on April 30 Schwartz and Martin arrived at the DOE's headquarters in the Forrestal building—a boxy and uninviting Brutalist building on Independence Avenue—Peña told them the staggering news that the message he would be delivering at the lab the next morning was that he was canceling AUI's contract outright, but leaving it in place pending the outcome of a competition. Schwartz was appalled; the plan he had brought along did precisely what the DOE wanted, with a fraction of the disruption, risk, and cost. Krebs's face was ashen, and would not look either Schwartz or Martin in the eye.[104] Afterward, Schwartz and Martin took off for the airport, Martin to go to Harvard, Schwartz to the lab. Aghast, Schwartz called Bond from the airport.

It was the first of four times that the lab would be blindsided, this time by the DOE. Peña would not wait for the "culture change" that Schwartz had prepared and promised, and would not permit AUI the chance to exploit the ISME's seven "opportunities for improvement" or the sixteen recommendations of the Bari report. AUI would get no opportunity to tell its story, as Krebs had urged, and would not be allowed to wait until 1999 when its current contract expired. While the DOE's own report blamed both itself and AUI, the agency would make only minor changes in its own oversight of the lab, and the only heads to roll would be lower ones.[105] As the options paper revealed, the DOE was taking the action principally for reasons unrelated to the tritium leak: to ease public and political pressure on the agency, to bolster its reputation, and to shock the other labs, as it was the first time that any contractor had been dismissed with an ongoing contract in place.

April 30 was also Samios's last day as director. It should have been celebratory; he was an eminent physicist who was greatly respected by the laboratory community, and had brilliantly rescued the lab from disaster when he first stepped into his position fifteen years previously. Not only that, the lab was hosting a party in the director's office to celebrate the signing of an innovative agreement with Japanese physicists to set up a center at BNL (RIKEN-BNL Research Center) to support early-career physicists. It was a seminal moment in international scientific collaboration: a center largely funded by another country on US soil to recruit and train scientists from around the world. RIKEN has continued to this day, and produced a large number of successful young scientists. The event had drawn eminent US and Japanese scientists and officials, who at the end of the day were partying in the director's office.

When Schwartz arrived back at the lab, he pulled Samios into another room and broke the news. The two sat stunned, gazing blankly at the walls, until someone appeared to summon Samios back to the party.

3 CHAOTIC SUMMER

On Long Island . . . so-called nonproblems are the worst problems to deal with.

—FRANK CRESCENZO

MAYDAY

Department of Energy secretaries rarely have time to visit national labs. When they do, lab officials pull out the stops. Two and a half years before, when then-DOE Secretary Hazel O'Leary visited Brookhaven, she toured the lab, addressed an overflow audience in Berkner, and left with souvenirs, including T-shirts sporting lab logos. When Secretary Peña arrived on May 1, employees assumed it would be the same deal: he'd come with the same canned DOE speech, and leave with the same prizes.

Peña quickly went off-script. He asked DOE rather than BNL officials to take him around and had a series of private meetings with DOE's Area Office officials. He lunched with them rather than lab staff, and closeted himself afterward. Peña then stunned lab officials (except the handful who had learned from Schwartz) with the news that he was terminating AUI's contract and shrinking the re-competition to six months so the new manager would be in place by November, but that he was leaving AUI in charge until then. The DOE's Area Office officials learned only at that moment why the April 29 press conference had been rescheduled.[1] "They chewed us up and spat us out," said DOE Deputy Site Manager Frank Crescenzo.[2]

Figure 3.1
DOE Secretary Peña (left) with John Wagoner, Jeanne Fox, and Tara O'Toole (right)

"We were shell-shocked," Bari recalled. "We had a sense things weren't going well, but didn't expect this."[3] Bond asked if AUI could rebid for the contract. Peña said it could, but that it would face "high hurdles." Peña spoke to lab employees for twenty minutes in the Berkner auditorium. He praised their work, but said that "the combination of confusion and mismanagement that has been occurring here over the years is going to end."

Peña and O'Toole then held a news conference attended by what seemed every media outlet in the region. "Termination of the contract," he said, "is a result of unresponsiveness on the part of AUI to address DOE needs and expectations of community relations and environment, safety and health stewardship."[4] He was, he continued, "very troubled and concerned [about] . . . the state of the relationship between the Lab and the community." To improve things, his agency would send two experienced DOE officials to the lab to replace Area Office Manager Carson Nealy: Hanford manager John Wagoner as executive manager of the local DOE group, and

Dean Helms from the Jefferson Lab to assist him.[5] Peña said he had asked Krebs to study the DOE's management of the lab; when the DOE rolled out its "Action Plan" a few weeks later, Schwartz was incensed that it was essentially his own Management Systems Improvement Plan.[6]

Peña also said he was asking the EPA to make an independent investigation of the lab's regulatory compliance, and that the EPA investigators would arrive Monday.[7] On the podium next to him was Jeanne M. Fox, the EPA's Regional Administrator (RA) for the New York area; her presence was designed to make Peña's action seem careful and methodical. In reality, her presence was anything but. The previous day, Fox had flown to Portland, Oregon, for a long-scheduled meeting of the EPA's national RAs. Around 5:00 p.m. on the West Coast (8:00 p.m. on the East) she got a call from the EPA's Washington headquarters—she wound up learning the news on a hotel pay phone—saying that the EPA, urgently requested by the DOE, was commanding her to return and to be at a press conference on Long Island the next morning. Fox's staff scrambled to book her on a red-eye from Portland to New York, picked her up at JFK airport with an extra set of clothes, and sped frantically down the Long Island Expressway as she changed. Fox gave her prepared remarks—that her staff had tested the drinking water and there was nothing to worry about—answered a few questions, then left to phone in to what remained of her national EPA gathering in Portland.

O'Toole followed Peña, adding that the ISME report had found that the lab's environmental safety and health practices required "significant management attention."[8] Like Peña, O'Toole charged the lab with being insufficiently sympathetic to the concerns of its neighbors. She declared that, for the recent lack of a few thousand dollars spent on two wells, millions were spent on cleanup.[9]

Not true. The DOE knew that installation of wells a year earlier would have affected only the cleanup's timeline, not its cost; only installation twelve years earlier would have saved money and time—and as the Bari report had just documented, the delay was partly due to the DOE's own prioritization procedures. Nor did O'Toole mention that the ISME report also severely criticized the DOE.

Peña and O'Toole tried to cast themselves as imposing justice on a chaotic and unsafe situation created by a reckless AUI. But many at the press conference were confused. The *New York Times* incorrectly referred to "the radioactive tritium that has reportedly contaminated the groundwater for miles around."[10] Other reporters thought that AUI was leaving immediately and that Wagoner now ran the lab, but AUI was left in charge until a contractor was selected.

Most of those at the press conference—including DOE and lab officials, employees, and reporters—appeared uneasy, and conversed nervously in small clusters. Only Wagoner, an experienced manager of toxic environments who flew in for the day from Washington State, was calm and relaxed, telling a *New York Times* reporter who tracked him down at the nearby Islip airport just before his flight back to Hanford that "from the data I've seen, the water is safe, absolutely."[11]

Afterward, Peña made a point of visiting neighbors of the lab, as he had when visiting Los Alamos a few weeks previously.

Schwartz, meanwhile, addressed employees. He had to choose his words carefully, and resorted to a sailing metaphor: that the day's events should not prevent the lab from plotting and beginning to sail a course so that, for whoever is at the lab's helm one or fifty years from then, "the sailing will be smoother."[12]

Peña, only six weeks in office and knowing little about either Brookhaven or the US national laboratory system, had taken a radical, unprecedented, and highly risky step in an unfamiliar and explosive environment. Could the lab's research continue? Would D'Amato back off? Would the activists quiet down or be reinvigorated? Could the HFBR restart? Could a satisfactory competition for such a huge contract be squeezed into six months? Would the firing paralyze the lab management? Would the lab even survive? Many in and out of the lab had cause to doubt every one of these questions. While all those questions caused unease in employees, foremost in the minds of most was, "Do I still have a job?"

A few weeks later, Peña ordered an Environmental Impact Statement (EIS) for the HFBR restart and said that his decision about the restart depended on its outcome.[13] While Environmental Impact Statements had been established by the National Environmental Policy Act (NEPA) of 1970, and involved public participation, they were not required for reactor restarts—and in fact the DOE had a strongly worded and enforced Reactor Restart Order stating that an EIS was not required in such cases. But Peña's initiating an EIS, and saying he would rely on the outcome, threw a speed bump into the controversial events and bought him time. It also meant that he would not fulfill his promise to decide on the restart by January 1998. Peña's statement now gave the activists a target, "Kill the EIS," and the lab scientists a new objective: "Protect the process."

Two weeks after the termination, a lead article of the *Science & Government Report*, an independent biweekly newsletter about US science policy, summarized the situation as follows: "Take the slovenly managerial habits of the Department of Energy. Add in laboratory administrators without a whit of public-relations sense, and let them loose on environmentally hysterical Long Island. Toss in a dose of Senator Alfonse D'Amato (R-NY), the undisputed sleasemeister of Capitol Hill, up for reelection next year. Then bring in a notably unqualified new DOE Secretary eager to show he's on top of the job."[14]

THE COVER OF *TIME*

By the freakiest of coincidences, the same week that Secretary Peña terminated AUI's contract, Brookhaven's research hit the cover of *Time*. The front of the May 5 issue showed a large fish whose open mouth was about to swallow a hook: "Sex Drugs Drinking Smoking: Scientists Are Discovering the Chemical Secret to How We Get Addicted . . . and How We Might Get Cured."[15] The story was about Brookhaven's research into the causes, effects, and cures of drug abuse, research just published in two extraordinary back-to-back issues of *Nature* magazine.[16]

Brookhaven's research into brain study and function had been one of its most successful programs for decades. In the 1950s, lab scientists had pioneered the development of positron emission tomography (PET) instrumentation for that purpose. In 1961, scientists in the lab's instrumentation division built the first single-plane PET scanner, nicknamed a "head-shrinker." For the next thirty-five years, scientists in Brookhaven's chemistry and medical departments had used PET scanning instruments to study the biochemical nature of brain processes. Early in 1997, Brookhaven purchased an advanced $1.7 million PET imaging device as part of a $2.2 million renovation of the lab's Research Center for Imaging and Neurosciences with the DOE and AUI collaborating in splitting the cost. At the dedication ceremony on May 6, it rained, and the ceremony had to be held inside the chemistry building with employees in the background, in the wake of Peña's announcement furiously preparing for the just-arrived EPA inspectors.

Two distinguished scientists led the Imaging and Neurosciences program, Joanna Fowler of the Chemistry Department and Nora Volkow of the Medical Department. Fowler, the head of Brookhaven's PET scan program, was a workaholic chemist who, in the 1970s, had co-developed fluorodeoxyglucose (FDG), one of the most important radiotracers in medicine, and spent two decades tirelessly perfecting its use. Thanks largely to her work, FDG is now used at PET centers all over the world to identify neurological and psychiatric disorders. Volkow, a psychiatrist and medical researcher, used FDG and Brookhaven's PET scanners to study the biochemistry of drug, alcohol, and tobacco use, as well as obesity, hyperactivity, and aging. She became chair of Brookhaven's Medical Department in 1996. (Profiles of her rarely fail to mention that she was Leon Trotsky's great-granddaughter, who grew up in the house where the famous anarchist was assassinated.) Much of her research involved tracing flows of dopamine in the brain when stimulated. She once had her own brain scanned, in the course of which her dopamine levels soared. "I'm addicted to work!!!" she beamed while taking her head out of the machine.

Meanwhile at the beginning of May, the Relativistic Heavy Ion Collider was taking shape north of the Imaging Research Center. In January, technicians successfully conducted a "sextant test" of a sixth of the accelerator ring two and a half miles in circumference, with its expected commissioning in 2000. At the Alternating Gradient Synchrotron (AGS), the lab's most powerful operating accelerator—revamped to become the injector for RHIC—experimenters were conducting several sets of important experiments. One consisted of tests with heavy ions, to train scientists in the new field of physics that RHIC would open up. The second set studied extremely rare events in proton collisions, work for which the AGS was uniquely poised because of upgrades to its luminosity, or beam intensity. Still a third important experiment created a beam of muons and measured a property known as the magnetic moment, whose precise value would provide a clue to the soundness of fundamental theories of high-energy physics.

Elsewhere in the lab researchers were developing advanced mammograms, high-performance cement, asbestos removal methods, sensing systems for antiterrorist efforts, and research into the causes of massive algae blooms known as brown tides, which destroy shellfish and had nearly destroyed Long Island's shellfish industry a decade before. Lab biophysicists built a model of a key Lyme disease protein for testing as an experimental vaccine.[17]

The tritium leak, and Peña's reaction, threatened all this work thanks to budget cuts, low morale, and canceled projects, not to mention more calls to close the lab. Already by mid-May, the cost of the tritium remediation had risen to $15.4 million, of which $11.2 million had come from the lab's research budget. The AUI trustees felt that the drain on funding and other DOE actions were meant to put pressure on them to decide not to restart the HFBR in order to rescue the DOE from its ugly predicament. "Such a scenario," according to the minutes of the June meeting, "would result in a great loss to the Nation as well as a significant loss to BNL, but it would also give the Office of Energy Research the outcome it desires without its having to make an unpopular decision."[18]

The National Institutes of Health withdrew a grant from BNL on the grounds that AUI, the legal entity through which such funds were transmitted to the lab, would not be in place.[19] Brookhaven's oceanography program—which included the lab's studies of the brown tides that periodically threatened Long Island's shellfish industry—was terminated, and the lab's program to treat Marshall Island victims of radiation exposure was taken away. The Protein Data Bank, founded at Brookhaven in 1971 and one of its great contributions to science, was transferred to Rutgers University. The PDB was and remains the most important open-access, digital data resource in biology and is required reading for educators and researchers needing to know protein structure and function, and its transfer was a major loss. Peter

Stephens, the chair of the NSLS Users' Executive Committee, told users that they needed to write to the DOE and congressional officials about the damage to research, saying that the DOE's action "may drastically change the funding and research environment of the laboratory, the site of first class research work during the past 50 years."

Peña's action in May made clear to scientists throughout the lab what Fowler had told her group two months previously: that if a few curies of tritium underground could make politicians and activists talk about closing the lab, so could routine scientific research. No serious research program was safe.

NOT ACCORDING TO PLAN

The DOE's plans went awry almost immediately. Peña's action did nothing to dampen community concerns or increase trust in his agency or the lab, and it only temporarily won over politicians. Furthermore, his justifications for the action began eroding within days.

Peña had said he made the contractor change partly because AUI, the previous manager, had lost the trust of the community. He had to rely more and more heavily on that justification as his others—the significance of the tritium leak and the supposed incompetence of AUI in handling environmental health and safety—were exposed as weaker and weaker as legacy environmental cleanup had progressed, though there would be some surprises.

Peña surely expected his EPA investigation to find something incriminating to help justify his drastic step. It didn't. The preliminary report, which Fox issued ten days later, listed infractions in things such as record-keeping, expiration dates, and lack of labels and aisle space in storage areas, and mentioned some hazardous waste leaks and spills, barrels containing VOCs, and a few other things out of compliance, but "none of these violations appear to present an imminent endangerment."[20] The ongoing EPA investigation would be only the first of several independent investigations to conclude that the lab posed no danger to environmental safety and health.

Humiliating, too, was the discovery that the DOE maintained 29 fuel pool facilities, some from the 1940s, most unlined and made of bare concrete, susceptible to leakage, poorly monitored, and only half of which had monitoring wells at all. This, together with the conclusions of the ISME and Bari reports, also severely weakened the DOE's charge that Brookhaven had significantly deviated from standard practices in handling its spent fuel pool.[21] Furthermore, a comparison between the DOE's own studies of the Los Alamos lab and Brookhaven on their management of environmental safety and health revealed that though the "results, conclusions, and overall ratings were similar," the DOE's response was drastically different.[22]

Still worse, key DOE officials, including O'Toole, showed that they were uninformed. At a meeting in DOE headquarters in Washington, when an activist mentioned Brookhaven's "other" reactor (meaning the Brookhaven Medical Research Reactor), O'Toole said dismissively, "Not so," and kept denying that the lab had a second. Grossman then pounced: "The DOE is, in fact, as confused and mismanaged as the lab itself."[23]

In Washington, Representative Sensenbrenner was skeptical of the need to remediate the plume and to terminate AUI's contract. He and fellow committee member George Brown summoned Peña to testify before the committee, and later said the DOE has "removed—rather than reinforced—roles, authority and accountability."[24] Sensenbrenner and Brown also requested an investigation by the General Accounting (now Accountability) Office (GAO). "Both Mr. Brown and I," Sensenbrenner wrote, "believe the decision to terminate Associated Universities Inc.'s (AUI) contract may have profound and precedential implications for the DOE Laboratory system and for DOE itself."[25] The GAO investigators would arrive at the lab in September. "General Accounting Office asked to fix the blame," *What's New* cynically commented.[26]

The DOE had rolled the top head at the lab by firing Schwartz on the day he took over as director, but its own head-rolling was minor. Nobody in its Washington headquarters lost their job. Wagoner and Helms had awkward encounters with DOE Area Office Manager Carson Nealy, whom they were replacing, for a week or so, but DOE headquarters ordered Nealy to move to an office at the Islip airport for a few months and forbade him from returning to the lab.

Arriving for his first official visit on May 13, Wagoner told lab officials that he felt sheepish being at BNL, for at Hanford, his home base, highly radioactive wastes were leaking from giant, single-lined underground tanks, some holding a million gallons—and it was next to impossible to stop them from continuing to leak. While Brookhaven had a leak of 15 or so curies of tritium that would never reach the lab's border, Hanford was dumping 6,000 curies of radioisotopes more long-lived than tritium *each year* into water supplies that fed the Columbia River, with 200 *million* more curies of yet-to-be removed waste.[27] Wagoner's second day ended abruptly. A chemical storage plant exploded at Hanford, blowing holes in the roof of the facility and exposing eight workers to toxic chemicals—and that lab's response was chaotic, revealing a lack of adequate planning for emergencies.[28] The DOE recalled Wagoner from Brookhaven and sent him back to Hanford. He didn't return for some time, and for the remaining several months flew back and forth across the country. During his days at Brookhaven his role included meeting with CWG members as well as Suffolk County officials; while he was gone, Dean Helms, a careful, conscientious, and cool-headed manager, effectively became the Area Office manager. Wagoner left after three months, at the beginning of August, when Helms took over.

Nealy, cooling his heels in the Islip office, was given unpleasant but necessary chores. When Peña finally learned that he was firing an organization whose performance the DOE had graded "excellent" just the previous year, he handed Nealy the task of telling Schwartz that the DOE was retroactively changing its performance grade to "marginal." This was despite the fact that Nealy's May 16 letter agreed that using the "goals, indicators, performance measures and metrics [in the DOE contract]" justified the "excellent" rating. The DOE's regrading, he stated, was going beyond "the mere application" of the procedure in the contract. Nealy was also given the job of correcting O'Toole's remark that earlier installation of monitoring wells would have saved millions.[29] After several weeks of such menial labor Nealy was transferred to the DOE's Chicago office, from which he retired after a few years.

Peña's action threatened the existence of AUI, whose original and primary reason for being was Brookhaven. Why hadn't Peña simply "fired a shot across the bow of the BNL flagship," asked AUI Board Chairman Paul Martin, "instead of firing a direct hit on its command station"?[30] At an emergency meeting on May 9, their first since the firing, the trustees were outraged. They noted that, from a scientific and regulatory perspective, the leak was not a health hazard. The DOE was imposing standards developed for weapons labs with major environmental issues on a basic research lab with comparatively trivial ones. It was absurd to import someone from Hanford, which was dealing with spills of thousands of curies, to address a leak of 15! Why use a howitzer to kill a flea? The remediation was not scientifically serious but an attempt to "give the community a sense of assurance,"[31] and it hadn't even done that. From an environmental safety and health perspective, the lab was even "probably in better shape now than it has ever been." Some trustees suggested "going political"[32] to reverse the termination—but that was a nonstarter given that they'd need to get political with Forbes and D'Amato. AUI was, a trustee remarked, exposed to the front end of the battering ram. The trustees made a Freedom of Information Act (FOIA) request to the DOE for documents about the decision to terminate them in case they decided to pursue legal action.[33]

Schwartz was in an awkward position, for he was both lab director and AUI president—and the DOE had approved him as the first and effectively fired him as the latter. It was not the lab but the AUI that Peña had fired, and it could be confusing whom Schwartz was representing when he spoke. Following the May trustee meeting, Schwartz, in his AUI president role, sent a blistering letter to Nealy that the decision was "based on bad faith, an abuse of discretion and constitutes a breach of contract," and asked that the DOE rescind the termination. Schwartz reminded Nealy of the ISME report's conclusion that the lab's actions in response to the leak "have been aggressive and appropriate" and that the DOE "shared responsibility for specific events as well as the overarching management."[34] He also argued that retroactively "regrading" the lab's performance was improper. "[T]he termination stands," Nealy wrote back, as instructed.[35] Schwartz sparred futilely with the DOE for months, via Nealy as go-between.[36]

Another DOE oversight was that Peña had not said he would require the new contractor to rehire current employees, in particular, tenured scientists, who had a commitment from AUI to support them for six months after the contract termination. To shield itself from this obligation, given the DOE's goal of a new contractor by November, AUI needed to give all tenured lab employees a termination notice immediately. Schwartz met with senior scientists on May 27 in Berkner and showed them a draft "Notice of Termination of AUI Scientific Staff Status," which became known as the "Pink Slip" or "We're All Fired" memo.[37] He explained that it was merely a legal requirement and that the scientists would probably be hired by the new contractor. But the word "probably" was no consolation: AUI, in fact, would be giving them notice.

When Schwartz finished speaking, he asked for questions. The first was from Robert Bari, who demanded that Schwartz resign. The next two speakers demanded the same. The meeting went downhill from there. Questioner after questioner hammered Schwartz for conflict of interest: as lab director he was charged with looking after the employees' welfare, but as president of AUI he was charged with looking after AUI's corporate interests. Who was he representing? The scientists were bitter that they had no real spokesperson. One excoriated Helms and other DOE officials for catering to the extremists and politicians and for not saying firmly and explicitly that the tritium in the ground was no health threat. When Helms said it was important to be sensitive to community concerns, one speaker exploded and noted that some people have charged that the lab has space aliens—"Should we not tell them that there are no space aliens here because we respect them?" The meeting, according to *What's New*, "dissolved into turmoil."[38] Several scientists declared they would seek to go elsewhere.[39] "It was traumatic," recalled Schwartz. "To the scientists it looked like we were running out on them."[40] The "Pink Slip" memo was never sent, but the damage was done.

The day after the "we're all fired" meeting Schwartz knew he needed to step down as interim lab director, and the Search Committee began scrambling to find another interim director.

In late May, DOE official Steven Silbergleid arrived to explain to employees details of the contractor selection process. The employees barely paid attention, turning it into another occasion to vent. Why are we going through this, asked one, when the tiny amount of tritium in the ground has *nothing* to do with public safety? The leak should have been detected, yes, but to claim that it is a health problem is a *farce* and the DOE's actions in providing public water hookups have just encouraged the extremists. Why doesn't the DOE admit that what is happening at BNL is a complete *nonproblem*?

"From a health impact, you are correct," Crescenzo responded. "But it has been my experience on Long Island that so-called nonproblems are the worst problems to deal with. The public, whether people want to accept this or not, is very upset with BNL and the DOE right now, and you can't deal with it by simply telling them that they're wrong."

The questioner was unsatisfied. Yes you can. "One tells them that they're wrong many times. The Secretary could do that . . . he should take the lead . . . we need courage . . . leadership." That response was revealing. The questioner expected that what had worked in the past—find somebody with enough clout to tell people that they are wrong and repeat it loudly enough until they get the message—would continue to work. But that this was the expectation was part of the problem.

The DOE's firing of AUI backfired, a lab scientist bluntly wrote Peña. Like the free water hookups, the new action has "only inflamed our relations with our neighbors while increasing their anxieties." Many in the community see the contract termination "as confirmation of their worst fears: something truly terrible must have been going on to warrant such an unprecedented action." Stop the "grandiose gestures" and come visit the lab again and be honest with the community about the lack of danger that the lab poses to the community.[41]

Peña's action alarmed national and international scientific media. Its eventual outcome, *Nature* editorialized, was likely to be more obstacles to effective research due to "imposing yet more oversight, review and regulation on the Department of Energy's laboratories." Citing the DOE's own 1995 review of its often byzantine oversight of national laboratories, *Nature* continued, "That is the last thing they need."[42]

AUI's sin was not corruption or incompetence, a scientist from the California Institute of Technology declared. "It was the ultimate form of modern-day malfeasance: AUI failed in public relations" and in not being "sufficiently sympathetic to rumors and the irrational concerns of some neighbors." The scientist continued:

At one level, there is nothing new here. Better societies have always consisted of a large body of decent, reasonable people who tolerate a few difficult ones—the credulous, the lunatic and the malicious. What is new is that the difficult ones—for whom perception equals truth—appear to have gained the attention and respect of our government. For whatever inappropriate reason, a senator and a congressman now endorse their bizarre scenarios. The claims of any group—say, Proctologists against Plutonium—appear to carry as much weight with government as do reasoned responses from major science organizations. And not suffering these fools gladly may be grounds for reprimand and dismissal. It's a shameful situation.

Department of Energy officials call for the lab scientists to be more sympathetic to the concerns of its neighbors, the author continued. Really? "Should the American Institute of Physics host conferences for flatearthers? Should *Physics Today* begin publishing accounts of survivors of abductions by aliens? Or let us have the American Physical Society finally create a division of astrology and necromancy—to show that we care. I am sure that the Department of Energy would be impressed."[43]

FRIENDS OF BROOKHAVEN

The day after the termination, on May 2, Bill Graves, a research scientist at the National Synchrotron Light Source, called his friend in the Biology Department, John Shanklin, who had been disturbed by the attacks on the lab, and who had written letters to newspapers.

"I gotta talk to somebody," Graves said.

Shanklin asked if the swallows were out yet. Graves said he didn't know.

"Let's find out," Shanklin suggested.

After hours, the two, joined by physicist Ben Ocko, walked along the headwaters of the Peconic River near the old Camp Upton grenade range, and shared feelings of helplessness. The DOE had lopped off the lab's leadership, activists were spreading false information that the media were amplifying, and politicians were shamelessly exploiting it all.

The trio were frustrated at how little their voices were being heard. Week after week, militant activists turned up at meetings—some dressed in lurid costumes—stated their positions in over-the-top language, and instead of staying to discuss with the scientists they would leave to be interviewed by reporters. The activists were more colorful than the scientists, and the ones who got standing that night on the news.

Over the next few days, Graves, Ocko, and Shanklin canvassed friends to see if any wanted to create an informal action group. They shared "a sense of outrage," Shanklin put it, over the difference between "the way *Newsday*, politicians and activists reported news about the lab" and their own experiences. Shanklin, who had been at the lab half a dozen years, found it "a place where intellectual contribution was valued above numbers of published papers. A place where, if someone was ill, people got together to help the family. A place where newcomers from the outside were welcomed with information, help and support. In short, it just seemed like a place that had values that really mattered."[44]

Bioinformatics specialist Sean McCorkle was interested. He had attended a Forbes town hall and had been terrified. "I thought, 'This is what it must have felt like to be a fourteen-year-old girl charged with witchcraft in the 1600s,'" he said. "People were saying that Brookhaven had nuclear weapons and was carrying on human radiation experiments and was poisoning the water. It was like an *X Files* episode—but they believed it was real. When I came home I had to have a couple of beers to calm down." In McCorkle's high school days in Maryland he had read science books with pictures of Brookhaven and its discoveries. "I assumed everyone knew about the lab," he said. "I found it astounding that people who lived a mile away didn't know much about it, and shocking that some people thought it was evil."

Joanna Fowler, who was still running her "Looking for Trouble" group, was also interested. "We were in a place we'd never been," she recalled. "We were realizing how precious the lab was, the freedom we had in it, and how important that was to young scientists. I loved the lab, and I was afraid we'd lose it."[45]

Others who were interested included Jean Jordan-Sweet, an IBM employee stationed at the NSLS; materials scientist Jim Hurst; Ed Kaplan of the Department of Applied Sciences; atmospheric scientist Steven Schwartz (no relation to Lyle Schwartz); Frank Marotta, who was both a lawyer and the lab's emergency services manager; and Jennifer O'Connor of the Reactor Division. Phil Pizzo, deputy head of the International Brotherhood of Electrical Workers (IBEW) at the lab, initially attended as an observer but soon became an active member, interfacing with the union.

The members named their informal group Friends of Brookhaven (FOB). It was a collection of Brookhaven employees who ranged from senior scientists to managerial and clerical staff, and a union representative. As Shanklin commented, "It really was a cross section of the lab, and quickly became a cohesive group in which everyone's voice was considered equal, which generated a unique feeling of trust and community."[46] With no experience in politics, media, or community relations, though, they needed advice. Graves knew a prominent Suffolk County attorney experienced in environmental issues, and half a dozen FOB members spent three hours with him on a Saturday afternoon.

It was not pleasant. The lawyer "painted a bleak picture," read Schwartz's notes, telling them just how steep an uphill battle they faced. The lab is all but helpless against the combined onslaught of activists, politicians, and the media, the lawyer said, because "there is no political constituency for the lab . . . there is no clear articulation of why the laboratory needs to continue to exist." Peña is likely to find closing the lab "the easy course." The only way to stop him is to make the course of action more difficult, making it clear that "dismantling the laboratory would be like dismantling the Louvre because there is a leak in the roof." Peña must be portrayed as "the Philistine pulling down the Temple of Science."

The lawyer continued by advising the group to form a professional-style organization with a president, secretary, and treasurer, whose members would do slick outreach and community involvement. They would have to paint a picture of the lab and Long Island as working together to make the area a uniquely attractive place to live, and to come up with catchy slogans such as "Synergism of the Beaches and the Brains." The group should also take legal action against the Department of Energy, which would cost $130,000 for the brief, and $400,000 if it came to trial.[47]

The lawyer, in short, told them that they had to behave like the activists. "It creeped us out," said Fowler. The members of the group became a not-for-profit—501(c)3—corporation, and tried a few of the lawyer's suggestions, such as handing out stuffed animals at a children's hospital. There's a picture of Fowler and three other FOB members standing in front of St. Charles Hospital in Port Jefferson with stuffed bunnies and a person in a bunny suit. Nobody looks comfortable, not even the bunnies. "It smacked of using the children," Fowler said. "It wasn't our lane."[48]

Among the Friends of Brookhaven's fears was the looming competition for a new contractor. The group was afraid that the DOE wanted to install Raytheon or some other large for-profit corporation friendly to D'Amato, whose only interest would be in exploiting the lab's resources, and they decided that if the activists could use petitions, so could they. The next week, in their first collective act, FOB members collected 1,100 signatures on a petition, sent to Wagoner, with a set of demands for the contract competition. "We believe that the unquestioned scientific excellence of Brookhaven National Laboratory is founded on an independent academic atmosphere," it began. It asked for "commitment to multidisciplinary scientific research excellence," "a nonprofit academic operator," and "continuation of scientific tenure system."[49]

FOB members soon learned that it was not easy to write letters that newspapers would publish, and their initial, unsuccessful attempts were on the long and abrasive side. Shanklin wrote a lengthy, heartfelt, and carefully researched letter to the *New York Times* after it ran a story that alleged the tritium plume endangered the community. I'm not saying that the leak is insignificant, Shanklin wrote, but radiation is such an emotional topic that when the *Times*, *Newsday*, and other newspapers report the activists' claims and bury the evidence of what the harm really is, it terrifies members of the community, injures scientists, and wrongfully damages the institution.

Scientists think numbers matter, Shanklin continued. Numbers spell the difference between something being safe and dangerous. Not feelings or perceptions or emotions—numbers. When it came to Brookhaven's tritium leak, the press and the politicians aren't adding them up. "When I have told our politicians the numbers, they tell me numbers aren't important," he wrote, "but if I ask whether they ever send $700 in response to a $70 phone bill, numbers apparently start to mean something."[50] People keep saying there's a "pool of tritium" under the reactor; how much is that? A cupful? A swimming pool full?

To find out, Shanklin continued, he sought out a medical supply company, browsed its catalogue, and found that they sell tritium to scientists for research purposes. He checked to see how much it would be if he ordered an amount equal to the five curies in the HFBR's leak. The five curies, he discovered—using all caps for emphasis—"would come in ONE MILLI-LITER of water . . . less than one half of a TEASPOON . . . the amount in approximately 200 LUMINOUS WATCHES." He continued, "And, if you put one piece of Xerox paper between you and the water, no radiation could pass through because it is so weakly radioactive."

"In truth," Shanklin continued, "there is a pool of water that has some rather dilute tritium in it." Furthermore, for people to be harmed "they first have to be exposed—and the DOE, SCWA, NY State health department, and the EPA have all said again and again that there is no EXPOSURE therefore can be NO HEALTH EFFECTS." Shanklin concluded, "[C]overage of this situation is really out of control."[51] His letter was not published.[52]

The FOB members arranged to meet with *Newsday* editors at the newspaper's Melville office to talk about publishing letters from scientists, its coverage of the lab, and to complain about what they felt were often unnecessarily scary headlines—such as "Nuke Lab," "Ground Zero," and "Meltdown on Long Island"—given to otherwise reasonable articles, and about the often accompanying scary photos. Other complaints involved wording: *Newsday* articles had called the tritium "lethal"; said that it was "laced" or "lurking in" the groundwater, and that the lab had not "revealed" or "disclosed" the tritium leak, as if it had been kept secret and not reported as soon as the test results were verified. The FOB members complained about the photos and images. *Newsday*, for instance, sometimes ran a map of the plumes coming from the lab site, the drawing looking like a malevolent seven-fingered devil hand about to seize the unsuspecting communities to the south. But, Shanklin and company protested, if they had run a similar picture of the hundreds of chemical plumes dotting Long Island left by its industries and factories—especially Grumman at its various locations, including nearby Calverton—the lab's plumes would be difficult to find.

The group spent days designing and writing a trifold flyer about themselves: "We care about our neighbors, our environment, and our science." They wrote other flyers to combat disinformation spread by the activists, to show the scientific research of the lab, and to explain the tritium leak in a reasonable way, printing thousands to hand out at public meetings. They spoke when they could with local officials. "I (and others) had plenty of contacts with the town, county, and state," recalled Marotta, "but they were afraid to talk to us on the record because we were 'toxic.'" But they developed contacts that could mediate between them and the local officials who wanted to stay out of the loop. "I found these behind-the-scenes discussions with people with whom I had credibility helpful because they freely told me their perceptions and suggested ways to help . . . off the record."[53]

The FOB members created displays for the lab's Visitors' Day, judged science fairs, gave out small cash award scholarships, and organized an effort to plant ten thousand seedlings on the lab's grounds. They put together a talk on the benefits of nuclear medicine in blood imaging and in diagnosing and treating cancer, which had benefited millions of people, in what the FOB members jokingly referred to as the "Joe six-pack" presentation.[54] They created handouts with easily understandable explanations of background radiation, contrasting it to the comparatively minuscule amount of radiation given off by BNL. "We lifted each other up," Fowler said.

THE TRANSPARENCY PARADOX

Going forward, the DOE would need to justify its termination of AUI's contract solely on claims of the loss of community trust. But far from calming the metaphorical waters, Peña's action did not much help to rebuild that either, and instead energized the more extreme activists and led to yet more calls to close the lab. "What the planned change in management would amount to?" Grossman asked rhetorically, dismissing the competition to replace AUI as "Zero or worse."[55]

The remediation plan—to siphon water below drinking water standards and send it back upstream, which was tested, approved, and implemented in May—also did not help calm the metaphorical waters. Caldicott called it "dangerous" for it would be "raining tritium over eastern Long Island," while various scientists called it "not good science" and a pointless public relations stunt.[56] Why had the DOE done it? After interviewing DOE officials, the *New York Times* concluded that the plan had been "rushed for political reasons—by Senator D'Amato to protect an important base of support before his re-election campaign and by the Energy Department." When an EPA official tried to explain to the senator the technical nuances of remediation options, and the lack of scientific necessity for them, the senator replied that "he was not a detail person and saw only one alternative, and that was to pump the plume. . . . People are entitled to action."[57]

Paradoxically, too, the DOE's insistence on transparency about the new facts that were being discovered about the groundwater contamination only sewed more fear and distrust. The reason had to do with the nature of the facts.

Howard Schneider, managing editor of *Newsday*, liked to tell his journalism students about a newspaper article, published shortly after Hurricane Katrina, incorrectly stating that National Guard troops had found 40 bodies in a freezer at the Superdome. It turned out that the journalist had not actually seen the freezer and had relied on a rumor. Schneider instructed his students to "Open the freezer!"—that is, verify information firsthand. It was a good lesson—an inspiring and instructive story. But—as in the Brookhaven context—what if what's in the freezer, unlike bodies, can be described only with numbers in unfamiliar units? The only way to find it out is to trust the experts who measured and interpreted the numbers. For many in Brookhaven's neighborhood, that trust was missing.

"I am giving Brookhaven every opportunity to be open," said antinuclear activist Dawn Triche when the leak was first discovered, but "a lot of people just don't believe what they say."[58] In the new transparency, every time a new finding, however trivial, was announced, it seemed to confirm the lab's incompetence in not having discovered it earlier. "They go around with this regular traveling road show to all the communities saying, 'Don't worry,'" Sarah Nuccio, another antinuclear activist told the *New York Times*, "but then strontium-90 shows up—lost in the system. That hardly builds anybody's confidence."[59]

If an independent institution or study reached the conclusion that the groundwater was, in fact, safe, the more extreme activists attacked those institutions and studies as incapable, corrupt, or untrustworthy. Grossman ridiculed institutions including the Suffolk County Water Authority and the Long Island Groundwater Research Institute. Maniscalco accused officials of the Suffolk County Health Department of lies and cover-ups.[60] After some LIPC members said that the HFBR did not pose a threat to the community, Helen Caldicott sought to have the IRS question the organization's tax-exempt status.[61]

When lab scientists addressed public meetings, their focus on evidence, use of phrases like "vanishingly unlikely," and references to sources of uncertainty seemed to undermine their message and were no match for Maniscalco's doomsday predictions, Smith's dead-certain pronouncements, or Caldicott's apocalyptic imagery. At one community meeting, during a scientist's talk, Maniscalco suddenly leapt onto a table dressed as a skeleton, screamed, and left, trailed by reporters and cameramen; he appeared on the nightly news.

The lab's media relations staff were easily judged untrustworthy. *Newsday* characterized the lab's public relations efforts as "damage control"[62] and published a letter written in response to the article that claimed that the lab spent more money on public relations than cleaning up the environment and that the lab wanted to "supply the public with cancer and then sell us a cure."[63] Rowe replied with a letter pointing out that each national lab's Office of Environmental Restoration is required to carry out community relations activities and that the amount Brookhaven was devoting to them was only about 5 percent of its Superfund budget.[64] Grossman wrote: "PR. That's what it's all about to Brookhaven National Laboratory."[65]

The only ones who might have been able to take steps toward establishing trusting relationships were Mannhaupt and Essel, who had persisted (now without pay) in Community Working Group activities. But having once taken AUI's money, they were forever tarred. When Pannullo and Essel were quoted in *Newsday* as saying that D'Amato and Forbes were exploiting the tritium issue for political gain, hoping to burnish their environmental credentials, activists again accused them of having sold out. It seemed hopeless. "The politicians created so much momentum that now people are scared [of tritium]," Mannhaupt told *Newsday*.[66] "I felt like I was getting PTSD,"[67] recalls Essel.

National and international scientific magazines including *Science* and *Nature* regularly mentioned that the plume was not a health risk, and their articles tended to focus on issues such as the fate of the HFBR. Local papers such as *Suffolk Life*, the *East Hampton Star*, and the *Southampton Press*, for which Grossman wrote, generally gave the activists the most attention in often tabloid-like stories.

At the time, Long Island was a land of tabloids and tabloid fodder. Anyone who lived on Long Island in the 1990s remembers Amy Fisher, a seventeen-year-old high school senior who showed up at the house of her married older lover Joey Buttafuoco and shot his wife in the face. "Long Island Lolita," proclaimed the tabloids, which dragged the story on for months. Several non-BNL events in 1997 inspired some self-reflection within the media. Just after midnight, on August 31, 1997, Princess Diana, her lover, and their driver were killed in an automobile accident in Paris while trying to evade reporters hounding them, and the *Village Voice* commented, "We Killed Diana." The unfolding Clinton–Lewinsky scandal also made some reporters meditate on how they were treating sensational material involving politicians. Yet the 1990s remained, in the words of *Vanity Fair* magazine, the "Tabloid Decade."[68]

The media's often tabloid-like coverage of the lab included a *Long Island Voice* article accusing Fowler and Volkow of promoting cocaine addiction. "Coke Fiends Cruise LIE in Limos" was the headline; "The sweetest deal around" for coke heads "is at Brookhaven National Laboratory." Coke-heads are driven to the lab and back, get lunch, watch movies, and receive an honorarium—"free, government-sponsored coke, $100, limo ride."[69] Never mind that Volkow's experimental subjects received only small amounts of the drug, that it was not enough to get high, or that the research was helping find ways to treat and prevent addiction.

Following a practice that might be called "misleading magnification," reporters often focused on one detail and omitted the context. *Newsday*'s story "Bucket of Trouble," on the early spent fuel leak test, was an example;

including the context would not excuse the lab for not discovering the leak, but would have made the lab seem less incompetent and uncaring. In *Newsday*'s story "Coffee Mugs, but No Well,"[70] the lead photo was of a coffee cup with the words "Team Safety" on it, captioned "A coffee cup that got priority funding while a detection well didn't." Contextless, the picture and caption suggest a horrific disregard for safety. Here's the backstory: in the early 1990s, Brookhaven management started an employee safety program called "Team Safety." It was an incentive program, common in large organizations, in which employees are divided into teams that compete for prizes for the lowest number of accidents, injuries, lost work days, and so forth. At Brookhaven the prizes included fire extinguishers, gift certificates, and coffee cups printed with the words, "Team Safety." In March 1997, *Newsday* reporters, investigating why the promised spent fuel pool monitoring wells had not been installed on time, found out about the prioritization system and the rankings. A *Newsday* photographer then asked to see the safety projects and took pictures of a few innovative safety devices installed on big industrial machinery, including devices designed to protect arms and legs of operators. Knowing what the *Newsday* reporters were really after, Rowe steered them away from areas where coffee cups might be found. But the photographer eventually found the "Team Safety" cup he was seeking. The cup photo was more striking than one of limb-saving machinery, and the headline was incriminating: "Coffee Mugs, but No Well: Brookhaven Lab Panel Rated Gifts Higher Than Detection."

STAR

That summer, a powerful, well-funded, and celebrity-driven antinuclear group formed that would quickly overshadow all other civic groups who were taking on Brookhaven.

Bill Smith was the instigator. He called Alec Baldwin, one of the most prominent celebrities in the Hamptons, asking to meet him at a diner in Amagansett. Afterward, Baldwin wrote a letter to the *East Hampton Star*, entitled "Environmental Suicide." "Shut down B.N.L.'s reactors immediately," he demanded, for citizens have the right to "live free from a reckless toxifying government energy policy."[71] No matter that the HFBR was a research reactor unrelated to energy policy; people read the letter. Baldwin was also a master of scorching attacks. "They've never been forthcoming," he said later, referring to the lab. "They lied and lied and lied and covered up for decades. The whole lab is corrupt."[72]

That summer, Baldwin helped crystallize a powerful new organization based in East Hampton, called Standing for Truth About Radiation, or STAR. It was a commonplace acronym even on Long Island; it was also the acronym for the Save the Animals Rescue Foundation, and for the lab's on Solenoidal Tracker, a detector under construction at the RHIC accelerator. But all the money and celebrity—plus the fact that it was literally composed of stars—mentally pinned the acronym to the new activist organization.

Caldicott and others began going door-to-door in the wealthy Hamptons enclaves seeking support for the new organization, and it was soon packed with wealthy East Hampton inhabitants and celebrities. One benefactor was David Friedson, a wealthy Florida businessman who lived in Coconut Grove, Florida, and Manhattan's West Side, and who docked his yacht in Sag Harbor near East Hampton. Friedson, who became the president of STAR's board of directors, was the CEO and chairman of a $300 million-per-year consumer products corporation that, thanks in part to Jay Gould's participation as an expert witness, had received a huge antitrust and patent infringement settlement with Phillips, the multinational conglomerate. Still another prominent East Hampton resident who helped raise money was Ron Delsener, the flamboyant rock concert promoter.

For legal advice STAR approached Scott Cullen, another East Hampton resident from a prominent East End family that had founded the King Kullen supermarket chain. His grandparents had worked for the lab—his grandfather on the security force, his grandmother in computer processing—and he had swum in the lab's pool while growing up. But he had come to experience the tension, resentment, and distrust of the lab's neighbors toward the Brookhaven scientists. "There was a reason some people felt, 'Those fancy scientists—they don't care about us.' That perception in some communities was very real—and very raw at that point." In 1997, just out of law school, he had accepted another job elsewhere, "but this was so intriguing that I bailed on the other job and jumped right in."[73]

Another Bill Smith recruit was Jan Schlichtmann, an environmental lawyer-celebrity thanks to *A Civil Action*, which had won the National Book Circle Critics Award for nonfiction, with filming in process for the motion picture. Smith had read the book and knew of Schlichtmann's reputation as the go-to legal expert on cancer and groundwater contamination.

From the information he gleaned from STAR members, Schlichtmann saw similar issues at play on Long Island as at Woburn and Toms River. "I am sympathetic with people who feel powerless and abused," he recalled, and he understood their lack of trust in the reassurances that such people tend to receive. "The scientists' focus is narrow. Their reaction is, 'the contamination exists, but not that much, paddling in an open canoe is far more dangerous to your health.' They have an economic interest in there not being a problem. Meanwhile, there are families in the community, mothers get pregnant who are worried about their children, there's always disease in the area—does the disease just happen, or is there a contaminated source impacting our health? These questions can't be answered easily, and you get stories and piecemeal information." Schlichtmann had become a strong advocate of beginning with fact-finding research, and assumed that STAR's first undertaking would be to use scientific studies to determine whether there existed in cancer incidences the patterns that one would expect from particular exposures to particular toxins in that kind of environment. "You look at facts as they are not as you wish them to be." STAR, he told a reporter, is dedicated to fighting "the only true enemy that all of us have: our ignorance."[74]

Ever the dogged idealist as he had been portrayed in the book and movie, Schlichtmann was elated at the opportunity to help organize a group that was not on the verge of bankruptcy, whose members had yachts and beachfront mansions, and that would not require him to take out a second mortgage on his home and sell his clothes. "They had money, political and media influence, and were relentless," Schlichtmann recalled. "They had several orders of magnitude more power than your ordinary citizens group." He was also elated to be working, too, not against an all but impregnable company like

W. R. Grace and Ciba Geigy, but against an opponent—the lab—that looked like "it couldn't tie its shoes."

Until this point, most of the civic and community groups interacting with the lab had focused on groundwater. STAR, however, was explicitly an antinuclear group whose aim was to close the lab's reactors, and it realized that its best line of attack would be to tie up the EIS process.

Meanwhile, activists staged demonstrations that were combinations of protests and popular entertainment. July 4 happened to be the day of the landing on Mars of the Pathfinder robotic spacecraft, an event avidly followed by millions of people on television and the internet as a celebration of science and technology. It was also a day that antinuclear activists staged "Picnic and Protest" outside the lab gates; the flyer urged "Close the Reactors! Clean Up the Mess! Convert the Lab to Non-Nuclear Uses!" Signage included skull and crossbones images and mushroom clouds. Slogans were about "50 Years of Contamination" and saving children. The demonstrators planted a flower garden at the lab entrance to commemorate the Chernobyl disaster.

Brookhaven's scientists continued to scoff at such people and their antics—at least until the summer of 1997. For them, the antinuclear activists were living in an alternate universe where the only trustworthy evidence was that supporting their convictions no matter how unreliable the information or its source. It seemed to go without saying that their statements—easily refutable if one bothered to check—and over-the-top imagery delegitimized them and their cause, making it seem like there was no point paying attention to it. A visiting scientist recalled leaving the lab for lunch one day and driving past a group of activists protesting the HFBR. One of them was dressed as Jesus and carried a cross. "I said to my companion, 'Look at them! Who would take them seriously?' How wrong we were!"

FISHBOWLED

Peña's action, the activists' responses, and media coverage made Brookhaven National Laboratory a "fishbowl" where every event, big or small, was under a microscope, and the lab's every misdeed a headline. In the 1990s, Brookhaven had been slowly on its way to being fishbowled; the full phase transition culminated with the tritium leak. The DOE's mandated press releases for each event generated breathless headlines and closer scrutiny. In addition, the uncertainty around the lab's future and the continual attacks on the lab from activists and media were distracting employees' focus.

Three such events happened within one week in early June, adding to the drumbeat of disturbing incidents at BNL and the impression that the lab was out of control.

In the first, on June 3, a researcher was preparing to irradiate a superconductor material at the Brookhaven Medical Research Reactor (BMRR), and tested a sample of Saran Wrap to see if he could wrap samples in it. Unbeknownst to him, one component of Saran Wrap is chlorine, which becomes radioactive by capturing a neutron and becoming radioactive Cl-38. He and four others received a minor exposure, and the Cl-38, which has a half-life of thirty-eight minutes, vanished quickly.[75] No radiation was released from the building. Under ordinary circumstances, the harmless incident would have been investigated and addressed internally. The EPA and Suffolk County officials who investigated called the incident minor; but again, a press release had to be issued.[76] *Newsday*'s headline erroneously shouted that Brookhaven had had another "leak." While the text of the article said that "there was no radiation leak from the building,"[77] that word in the title was the part that most readers saw. Karl Grossman again called for the lab's closing.

In a second incident five days later, a worker violated procedures by moving a dummy fuel element from a storage area to a shop, and a small particle of Co-60 got transferred to his clothing and another worker got another small particle on his shoe; both particles were detected when the employees passed through a radiation monitor. The incident was harmless and no radiation left the building. Schwartz was away from the lab, and Bond—knowing the inevitable DOE and media reaction—ordered an HFBR safety standdown, forcing operations to halt, and he chaired a meeting with defiant reactor staff. Given recent events, Bond said, the incident has raised some major concerns about worker safety. It is true that "we are under a microscope," and nobody has been seriously injured. Still, "procedures are being forgotten or short circuited." We need to stop, slow down, and regroup, for "if HFBR [is] to restart we need to prove we can operate responsibly." Employees at the meeting pushed back. DOE's overly aggressive schedule to remove and package the spent fuel in the pool, some complained, required shortcutting procedures on fuel processing. The accelerated schedule required workers to do things they were untrained for, and they had too little input on procedures. Finally, the workers protested to Bond that the radiation portal alarms sometimes gave false alarms, a kind of high-tech wolf-crying, adding to confusion, stress, and cynicism.[78]

The third safety event within a week occurred on June 11, when analysis of the soil around the now-empty tank discovered and reported in March showed that it had leaked water containing strontium-90.[79] Heavy elements such as strontium travel very slowly in groundwater, had moved only 45 feet away from the tank, and did not affect drinking water wells. But because strontium-90 is associated with fission and fallout and is a symbol of nuclear explosions and weaponry (especially after the finding that thanks to atmospheric tests of nuclear weapons it could be detected in baby teeth), it became a catalyst for antinuclear fury. One headline was "2nd Dangerous Isotope Cited in BNL Leak."[80]

Periodic new findings of issues related to current or past practices, and the almost-daily news releases, would continue to project the image of a lab in chaos. Many at the lab were slow to recognize that the plume challenged the lab's self-image as having matters under control. For those outside the lab, it suggested that the lab was mismanaged and the DOE failing to supervise it. The danger might lie not in the leak but in what it appeared to indicate about BNL and the DOE.

Then on June 20, in a very serious accident, a nonemployee construction worker was killed at a lab work site. He and another employee of the construction company were grading an area in a remote area for a sewage line upgrade. The worker was hit by a large payloader driven by his coworker when he was holding a leveling stick to help gauge whether the ground was level. Wagoner was on vacation, Schwartz was in Washington, and Helms had to deal with DOE headquarters. Bond went to the wake with trepidation, but the family expressed no animosity toward BNL. "I do not hold the laboratory responsible at all," the father told reporters.[81]

Peña ordered a DOE investigation, as required by DOE policy, but also a "stand-down" of the entire lab from Friday to Monday, meaning that while the gates were open no activity could take place other than meetings to discuss safety concerns and other workplace issues. The stand-down mandate implied that the construction worker's death was yet another piece of the same picture as the tritium leak, and so the action made it look as if BNL was unable to act responsibly.[82] The DOE's investigation, however, attributed the incident to lapsed judgment on the part of the two workers, and suggested changes in evaluating subcontractors, especially about previous performance on completed projects and the required eight safety submittals. Pointedly but carefully, the DOE also advised the lab to ensure subcontractors have compliant and enforceable substance abuse programs.[83]

The DOE's imposed lab stand-down did not make sense to many employees and visiting researchers. One lab scientist described it as like a restaurant hiring a contractor to renovate, a contractor employee having an accident—and the local authorities then closing down the restaurant but not the contractor. As part of the stand-down O'Toole met with employees and promised that she would pay for its costs. That sounded fair—but what she meant was paying hourly workers during the stand-down and not the huge expenses resulting from canceled meetings and experimental runs, additional travel and hotels for users, and the cost of returning empty-handed to home institutions.

When Bond wrote O'Toole asking her not to blindside the lab, she responded, "I try hard not to blindside anyone because I think it is always a disastrous strategy," and proposed a "frank interchange" of views.[84] Taking her up on the invitation, Bond protested that the DOE had ordered the stand-down without consultation, and voiced more general complaints about the DOE's overall treatment of BNL. Now outraged that Bond had questioned the DOE's judgment, O'Toole replied an hour and a half later. Far more is at stake here than the lab's safety and morale. "I also believe the future of the lab is at stake to some degree and that the BNL issues play in the country's willingness to support national labs generally," she wrote. "I am really angry (and have been for some time) about the arrogance and ignorance of the lab directors as a group (with notable exceptions) regarding

ESH mgmt, which I think is inexcusable and intellectually dishonest. . . . This is not just a screw job, not just political grandstanding, not just a media frenzy, not just a misunderstanding by an ignorant public. That the fatality was not just an unavoidable accident, that BNL's pollution problems are not necessarily the cost of doing business or the inevitable result of past practices or just an easy means whereby the town fathers can bilk the feds for water hookups."[85] At the end she acknowledged that there should have been a dialogue with the lab before the stand-down was declared, but said it would have been better if the lab had called it.

Later in August a construction worker from a commercial company accidentally backed his forklift into a shower head at a waste handling area, breaking the pipe and releasing clean water into a drain. This, too, required a press release.[86]

WEIRD TIMES

It was a weird summer of weird events.

That summer, Brookhaven employees had to drive through daily protests at the lab entrance with placard-carrying demonstrators telling them they were environmental criminals and murderers. Lab employees—whose cars had identifying stickers—had shop owners, gasoline attendants, and others accuse them of poisoning their water. Some neighbors demanded that the lab be closed.[87]

There were also episodes that, if found in works of fiction, would be considered unacceptably contrived attempts at irony.

One occurred a few days after Peña's firing of AUI and involved the DOE's Radiological Assistance Program (RAP) team stationed at Brookhaven that included BNL scientists.[88] The RAP team investigated incidents involving radioactive materials in 11 Northeastern and Mid-Atlantic states. On May 11, the team was called to Union, New Jersey, after a sixteen-year-old New Jersey boy, as a prank, took three tritium-powered exit signs from a dump, broke them open, and inhaled some tritium. (After the 1993 World Trade Center bombing, many public buildings had replaced their electric exit signs with self-illuminating, tritium-powered ones, each with 30 curies of tritium, or about twice the amount in the HFBR plume.) The boy's urine samples showed 28.6 million picocuries per liter, or almost twenty times the peak concentration in the HFBR plume. His total exposure was about 300 millirems, about the average level of background radiation in the United States. Doctors, correctly, told him that this was not a health hazard, and advised him to drink lots of water (beer is sometimes recommended, but the boy was underage) and pee it out. The New Jersey Board of Health did not issue a press release because there was no threat to his safety. The RAP team confiscated the broken signs, sealed them up, and stored them at BNL.

"Flushing the toilet would violate EPA standards," mocked Robert L. Park in his *What's New* comment, which referred not only to the absurdity of the discrepancy between this episode and the firestorm over Brookhaven but also to the overcautious EPA standards. "Don't touch that flush handle!"[89]

A few newspapers mentioned the incident but did not link it to what was happening at BNL.

A *Newsday* column about Brookhaven's environmental problems was headlined "Tritium's No Laughing Matter."[90] But to many lab employees laughter seemed the only way to cope. At a lunch table in the Berkner cafeteria one day, a group of scientists competed for the most outlandish idea to remove the plume. Absurdity was no disqualification. The most insanely expensive suggestion was to install liquid nitrogen wells to freeze the ground, excavate the ground block by block, and build huge refrigerated airplane hangars at the lab to store the pieces. That came in second. The winning entry, which earned its inventor a free dessert, was to install water fountains on the HFBR's lawn and have employees drink from them and pee off-site. This was not dangerous—the New Jersey youth had proven that—and cheaper and more effective than the solution the DOE was paying for. But these were inside jokes, letting off steam under pressure, and surely would have provoked horrified reactions had the press gotten wind of them.

The growing reputation of Brookhaven as a secretive and ominous place made it fascinating to one soon-to-be famous author:

Dear Brookhaven,

I am writing in hopes of arranging a personal tour of your facility. I am a published novelist with St. Martin's Press and am currently at work on a thriller placed in a lab similar to yours (either CERN, Brokhaven [*sic*], or Fermi, I haven't decided yet). Do you by chance have anyone on staff with affiliations to Phillips Exeter Academy, Amherst College, or MENSA who might have a few hours to show a fellow grad around? (Or anyone on staff at all with time enough to give a tour?!) Either way I would love to visit. When you get a moment, would you be kind enough to advise me as to the possibility of a tour?

Thanks for your time.

Sincerely,
Dan Brown[91]

Brown, later the author of the best-seller *The Da Vinci Code*, was working on a novel in which terrorists steal a canister of antimatter from a famous laboratory to make a bomb able to destroy entire cities. Brown never ended up visiting BNL, and ultimately set *Angels and Demons* at CERN, but BNL had been a contender.

An author who did visit and publish a science fiction book set at BNL was novelist (and physicist) Gregory Benford. In *Cosm*, Benford painted the lab as a place beleaguered by "Safety Nazis" who have to fend off "rabid anti-Lab folks, like Fish Unlimited." Physicist Alicia and her grad student Zak manage to skirt regulations at the lab to run an experiment at RHIC that creates one space-time wormhole that kills another graduate student, and then a bigger wormhole that destroys the entire accelerator.[92] This hardly helped BNL's image.

Every so often, Long Island's UFO enthusiasts would claim that the lab had captured, shot down, or communicated with aliens. Ominously, besides Brookhaven, AUI also ran the National Radio Astronomy Observatory, whose telescopes at Green Bank were at the heart of the start of the search for extraterrestrial intelligence. Some UFO believers were convinced that the lab's tritium came not from the reactor but from a failed attempt to study an extraterrestrial spacecraft driven by a "clean" nuclear fusion reactor, and that the tritium had leaked out when the scientists opened the spacecraft incorrectly. "How are we to know what was released or where it came from?" commented one of the believers. "All we have to go on is what they tell us, and I don't believe them."[93]

The chairman of the Long Island UFO network, John Ford, charged Brookhaven with hiding aliens and being the site of UFO launchings and crashes. A lab physicist, he insisted, had told him of a weapons test in which lab scientists had shot down a UFO, causing it to split apart, fall into Moriches Bay, and explode.[94] "We didn't shoot down a UFO," Rowe told a reporter. That August, Ford went on trial for "murder with radioactive elements in retaliation for UFO cover-up." One of his ideas had been to grind up some radium and mix it with garlic to serve intended victims of Italian descent. Ford was declared unfit to stand trial and sent to a psychiatric center, still convinced that Brookhaven National Laboratory was covering up. "The truth is out there," he assured a reporter.[95]

"Brookhaven Lab: The Truth Is in There" was the title of a spoof in *Hurricane Eye*, the newspaper of the Westhampton Beach High School. Calling themselves "agents x, y and z," a trio of students wrote that they had disguised themselves and snuck into the lab in search of the aliens, rats accelerated to warp speed, and mozzarella sticks fried in plutonium said to be hidden there. "To our disappointment, we found only humans." Declaring this a cover-up, the pranksters challenged scientists about their research, but were frustrated to find them working on cures for AIDS, Lyme disease, and brain and bone cancer; on treatments of Parkinson's disease; and on cleanups of pollution and preventing further contamination. The reporters encountered a woman who worked at the HFBR, Jennifer O'Connor, who told them that she had two pregnancies at the lab and two healthy children in the on-site daycare center, and that the lab posed no health threat to her children, lab employees or neighbors. Catching on to the gag, she told the students that she'd be interested in trying a plutonium-fried mozzarella stick if they found one. Disappointed, and still claiming to be convinced that all the lab's scientists were part of a conspiracy, the reporters removed their disguises and went to the cafeteria, still determined to find the mozzarella sticks.[96]

The cover of the hardcore porn magazine *Rage* promised an article on the inside about "Nuking Long Island." Sure enough, alongside "Cowgirl," "Crapshoot," and "Samurai Like It Hot" was Karl Grossman's contribution "Nuclear Genocide: How Brookhaven National Laboratory Is Killing People."[97] The article cited "facts" from Gould and community rage from Bill Smith, who was identified as living on Shelter Island "downwind" of BNL (it isn't), describing his ambition to shut down the Brookhaven reactors and then "move on and shut down all the nuclear death machines at all the other U.S. government nuclear facilities."[98]

When Shanklin learned of Grossman's *Rage* article he was incredulous, and asked his wife to pick him up a copy of the magazine on an upcoming trip in which she would be chaperoning schoolchildren to the American Museum of Natural History in New York City. "Do you have *Rage?*" she asked the owner of a magazine shop. The shopkeeper asked if she were sure she wanted it. "Yes," she said, and asked for a bag to keep it in so the children wouldn't see. He vanished into the back room, slammed it on the counter, and hissed at her: "You should be ashamed of yourself!"

Newspaper publisher Dan Rattiner, who had founded the Long Island weekly *Dan's Papers*, was known for his hoaxes and satires. In one issue that summer he wrote that he had found a time machine hidden in a lab closet, which he had used to discover that the lab was about to emit a cloud of "Xanozx Manathol," a deadly substance that will hover over all Long Island; meanwhile, 262 dead chickens will be found in a lab broom closet. The next year dinosaurs and pterodactyls will be spotted at the lab feeding on glowing green goo; they will feast on attack helicopters. A 40-foot spider will arrive and eat some but not all of the dinosaurs. A spaceship the size of Rhode Island will appear over the lab, and with a beam of light suck up the spider and remaining dinosaurs into its cargo bay. Three reporters will receive the Nobel Prize for Literature for *Slaughterhouse Six*, their book about the episode.[99] Rattiner found that his readers mostly didn't get that he was spoofing media coverage of the lab, and in the next week's issue laid it out explicitly: "Closing the lab would be, in short, probably the single biggest research disaster to ever take place in America."[100]

Weird, too, was the incongruity between events at Brookhaven and elsewhere. As *What's New* put it, "The miniscule tritium leak from the reactor at Brookhaven, which led to a change in management, is nothing compared to leaks at the Hanford reservation on the other side of the continent, where 67 of 177 storage tanks are leaking really bad stuff."[101] Savannah River had had releases of 2 million curies of tritium per year, and was struggling to get the amount below 100,000 curies. When a colleague sent Bond a report that Fermilab routinely released water containing 2,000,000 picocuries per liter he wrote back that there had to be a typo in that figure—an extra 0 or two. There wasn't. The DOE was fine with Fermilab's release of 2 million picocuries because of the difference in the type of soils; Fermilab's tritium pool was over a clay layer, while Brookhaven's pool of 30 curies was in a porous layer over an aquifer.

The lab held regular open houses for members of the community where scientists would explain tritium and its effects. Gunther and Rowe would bring cookies to the meetings, and found that some people refused to eat them out of fear that they were made with tritium-containing water. "I ate a lot of cookies," Gunther recalls.

DIRECTOR SEARCH(ES)

On June 6 two members of AUI's Director Search Committee went to Philadelphia to interview their leading candidate, David Moncton. Reflecting how seriously they were taking the search and their continuing estimation of their own prestige, the trustees chartered a plane to take two interviewers from Islip, the airport nearest Brookhaven, to Philadelphia and back, a flight of about an hour.

In principle, the forty-eight-year-old Moncton was an excellent choice.[102] He had a deep knowledge of Brookhaven, experience with neutron and light sources, a solid record as a manager of big-science projects, and was an associate laboratory director at Argonne National Laboratory, another multipurpose DOE facility. Career-wise, becoming a lab director was the natural next step up.

Two days later, on June 8, Schwartz met with Krebs to sound her out. Krebs said she was neither for nor against Moncton, and wanted to talk to him before responding. She told Schwartz not to have the AUI board vote on Moncton until she got back to him—and it was indeed the practice for a laboratory contractor not to propose a director without DOE approval. The DOE did not want to be in the position of having to justify why they vetoed someone.

The AUI trustees met three days later in Charlottesville, Virginia. One issue they debated was how to communicate the "injustice" the DOE had done to AUI: Op-eds? A white paper? Fact sheet? Timeline? Circulating the Bari report? They could not decide. The trustees also discussed whether to vote for Moncton as lab director. Though aware of Krebs's instruction, they resented what they saw as DOE bullying, and felt they had to take a stand against government interference and rebel, for only they knew how to make a scientific institution work. Not understanding the perilous political territory they were in, the trustees went ahead and voted to appoint Moncton the new interim director. Late on June 12, Moncton received a call from AUI head Paul Martin, chair of the board, about the decision.[103]

Moncton was skeptical. One obstacle was whether Argonne would allow Moncton to become director of one DOE lab when he was an employee of another. Moncton also had personal reservations: Was it wise, if AUI was in such trouble with the DOE? Finally, Moncton knew that AUI was unlikely to win the competition for the new contract, meaning that his new job was surely temporary. But he also did not want to insult Brookhaven, and told Martha Krebs that he would at least go through the motions.

One trustee was so annoyed that he put the DOE on the spot by leaking the Board's selection of Moncton to Park. The next morning—June 13—the lead item in Park's newsletter *What's New* directly quoted an action made in the AUI Board's closed meeting less than twenty-four hours previously.

Trustees of Associated Universities Inc. took action yesterday:

AUI PICKS A DIRECTOR FOR BROOKHAVEN NATIONAL LABORATORY.

The trustees of AUI hereby appoint David Moncton Director of Brookhaven National Laboratory effective on or about July 1 for the duration of the current contract.

The appointment is subject to DOE approval. Moncton, a condensed matter physicist and Fellow of the APS, is currently Associate Director of Argonne

National Laboratory. Lyle Schwartz, President of AUI, has been serving as interim Director (WN 2 May 97), but it was clear that the two roles involved a conflict of interest.[104]

Newsday followed up with an article quoting Moncton to the effect that he was considering the job.[105] Schwartz was forced to send an embarrassing walk-back of the news to lab employees on Monday.[106]

Krebs and the DOE went ballistic, and now adamantly opposed Moncton. After hours that day—June 13—Bond was sitting in his office when John Wagoner dropped by to ask Bond if he was willing to become interim director. Surprised, Bond told Wagoner that the DOE had rejected him once for the role, and mentioned the now-public fact that the AUI trustees had chosen Moncton. Wagoner told Bond bluntly, "Moncton will withdraw."

Five days later, still appearing to consider the offer, Moncton visited the lab, which announced that "Subject to DOE Approval, AUI Votes Moncton Director."[107] Meanwhile, Moncton told the AUI trustees that he was withdrawing, and made it public on June 24. The next day, Martin and Schwartz asked Bond to become interim director, and Krebs approved.

Bond took over on July 7, the first Monday after the July 4 weekend. The transition happened without fanfare: no speeches, no parties. The lab was, after all, only switching out one interim director for another—and there wasn't much practical difference, because Bond already had been acting the part when Schwartz went to and from Washington. The transition was privately marked: Schwartz got emails of thanks, Bond of good wishes.

Bond was now faced with persuading employees not to leave, with boosting morale, and with forging better relations with the press and the neighbors. He met with each of the lab's major departments and divisions. Three days after taking over, he and a few other lab officials met with the Mastic Beach Property Owners Association, an informational visit of the sort that the lab's administration had not made a practice of in the past. The event went well, and the organization supported the lab as the year progressed. The meeting—as well as subsequent meetings with many other outreach groups—was also an indication that the Suffolk County "community" was not monolithic but consisted of numerous groups with different attitudes toward the lab.

Bond also had to repair damage to lab programs. Volkow's path-breaking addiction research and cancer treatment programs had been hurt by false claims of danger, which had discouraged patients from coming on-site. Bond sent a letter to participants. "Because Brookhaven has been in the news a lot recently, some people have expressed concern about coming to the Lab to receive treatment or take part in a clinical study," he wrote. "I would like to personally reassure you that the local, state and regulatory agencies that protect public health have said there is no risk to anyone from visiting the Laboratory." Bond added, "If you are receiving radiation therapy here, please know that our thoughts are with you in your fight against cancer."[108]

STARTING THE RESTART

In early July the lab had pieces of good news that boded well for a reactor restart. The EPA audit that Peña initiated right after the termination of AUI's contract reported its Phase I findings, which followed the earlier preliminary findings in mid-May. The audit would continue for eight more months. The DOE had clearly expected major discoveries about environmental issues, but EPA Regional Administrator Fox announced that the findings concerned "poor operations maintenance and housekeeping." She added that "none of these violations pose an immediate threat to workers, the environment, or public health"; other EPA officials characterized the problems as no more extensive than at other facilities of comparable size and complexity as Brookhaven.[109]

More unexpected positive news arrived on July 14, when Forbes addressed a packed meeting of lab employees in Berkner and apologized for overdramatizing the lab's environmental issues. He said he lamented saying things that were "hurtful and, perhaps, over-characterizations" and that fed into media barrages against the lab. Still more promisingly, he indicated that he knew the HFBR was safe and that he was aware of its importance and supported its restart. "We don't want the reactor shut down because of . . . of folks who don't understand its importance and its safety—and it is a safe reactor . . . and has operated for many years without a problem."[110] Employees found Forbes's turnabout, and new acknowledgment and support, reassuring, if overdue.

The lab also began operating a groundwater treatment system of chemical plumes, largely cleaning and degreasing agents, at its southern boundary to prevent their off-site migration.[111]

A week after Forbes came to Brookhaven, the lab hosted an ISME workshop for the national labs. Virtually all the representatives reported that they had similar issues as BNL, but that none was getting the press that Brookhaven was.[112] There is no single recipe for achieving ISME, the meeting concluded; a lab's approach to ES&H must be tailored to the specific situation and balanced with the scientific program. The meeting ended without much conclusion.

Bond was alert to Krebs's April remark that the lab's directorship did not keep her informed. Anxious to improve relations with the DOE, two weeks after taking over, he went to Washington to update Krebs on lab activities, budget items such as transition costs, the fee for the new contractor, and a new tax agreement made with the Town of Brookhaven. Because Brookhaven National Laboratory was federal property, it was exempt from taxes, but in the next fiscal year, which began in October, the DOE would begin to give the town "Payment in Lieu of Taxes." The two also discussed the loss of the oceanography program and the Protein Data Bank, as well as DOE's reluctance to support the PET imaging program, which it was expecting the National Institutes of Health (NIH) to take on. The DOE began to phase out its support of the PET program, and while the NIH provided some support for the path-breaking program, it and ultimately the lab's entire medical research program would close in another decade. Ominously, Krebs also told Bond that the HFBR restart "is going to be a big issue," because of both its cost and the EIS process.

Then, on July 28, came an unexpected tritium finding when monitors of the water entering the sewage treatment plant picked up an unusual spike.[113] The lab, Suffolk County Health Department, the DOE Chicago Operations Office's Safeguards & Security Section, and the DOE Headquarters Office of Enforcement & Investigations all began separate investigations. The source was baffling. Samples were taken from wells all around the Peconic River, and all conceivable lab sources of tritium were investigated, including effluents from the reactor, spent fuel pool, Medical Research Reactor, and the AGS.[114]

The absorbent pads used to clean drains and sink traps were examined. The confiscated, broken-open tritium signs that the RAP team had taken from the New Jersey youth were located—but these had been sealed in bags and stored under a ventilation hood. All sources were ruled out. The likeliest possibility was a disgruntled employee who, to embarrass the lab, had tampered with the sewage system by dumping tritium into the sewage line after the monitors but before the sewage treatment plant. Such deliberate tampering had occurred in December 1996, when an employee's urine sample was found to have probably been intentionally spiked. Even though the total tritium release in the spike was well under the regulatory upper limit, the investigation into it would continue for months. Neither the BNL nor DOE investigations could come up with a better explanation, but neither could a culprit be identified. The conclusion of all the reports contained careful language such as that the spike might have been "intentional or inadvertent" or that deliberate action is a "possibility which has not been disproved." The DOE investigation on the tritium spike commented on "disillusionment with management" and that there was periodic talk of "testing the system," so it was possible that someone wanted to "get management by spiking tritium."

Forbes and D'Amato professed indignation that a definite cause could not be found, writing Peña demanding that the DOE send in a federal overseer and that an independent environmental manager be appointed at BNL. The two congressmen seemed uninformed that the DOE had already done so. D'Amato said, "It's like having the Keystone Kops in charge of a nuclear reactor or radioactive materials."[115]

Both proponents and opponents of the HFBR restart had welcomed Peña's decision to set in motion an Environmental Impact Statement process. Scientists who knew the reactor and wanted it restarted expected a positive result; furthermore, the proposed upgrade and increase to 60 megawatts would require an EIS, and this could be used as an opportunity to get the lengthy process out of the way. But antinuclear activists knew that an EIS could become an Achilles' heel, highly vulnerable to protests that would cause delays and increasing costs that could kill a project. Public hearings had doomed the Shoreham reactor, and the same might happen to the HFBR. It was a strategy of exhaustion: each delay would make an HFBR restart more expensive, and more of a frustrating distraction for the DOE.

The beginning of the EIS process galvanized the US science community, which was now acutely aware of just how close it was to losing its most important neutron beam facility. The HFBR was one of the three highest flux research reactors in the world, attracting hundreds of researchers each year. Its research programs included solid state and nuclear physics, structural biology, and chemistry. Five research instruments were vital to such research—small-angle scattering to study biological samples, triple-axis spectrometry to measure bonding and magnetic forces in solids and new materials, powder diffraction for identifying light elements in materials, single-crystal diffraction for studying crystal structures, and, in particular, location of hydrogen atoms to study surfaces and thin films. The HFBR provided a large fraction of the total US utilization of each of the five techniques in the previous year ranging from 14 percent to 50 percent. The HFBR was also an important source of radioisotopes used in DOE-sponsored programs for medical diagnosis and treatment.

In anticipation of the EIS process, Krebs asked the Department of Energy's Basic Energy Sciences Advisory Committee (BESAC) to recommend whether the HFBR should be restarted or abandoned based on certain parameters, such as that the review process would be completed in 15 months, with the restart in 1999.[116] A three-day BESAC meeting in Gaithersburg, Maryland, focusing on the HFBR was held from July 30 to August 1. Two and a half months of vigorous discussion and debate followed. In November, the committee sent Krebs a letter with their conclusion strongly endorsing a restart. Many of the HFBR's facilities were unique, the report said, and if the reactor were shut down these capabilities would be gone, including high-resolution crystallography and cold neutron research. The HFBR was also a valuable instrument for more routine studies given that the other US neutron facilities were oversubscribed. For instance, the HFBR provided half of the US beamline hours in single-crystal diffraction, and 43 percent in triple-axis spectrometry. The report concluded, "The Committee strongly recommends that the High Flux Beam Reactor at Brookhaven

National Laboratory be restarted, as soon as possible, to minimize the effect on neutron science research in the United States." Further, in what came as a shock to the DOE, the committee advised that, under certain conditions provided by the DOE, the HFBR "should restart at 30MW and move to 60MW in a timely manner."

At the end of the summer Forbes again changed his mind about the HFBR. On the eve of the Labor Day weekend, Friday, August 29, he wrote Peña, copying Krebs, "I restate my vehement opposition to any restart of the High Flux Beam Reactor."[117] He did not make the letter public, nor issue a public statement.

Just before Labor Day weekend Bond got a call from Forbes's chief aide, Fred Dombo, asking whether BNL could use money from the NIH. The answer was clearly yes, given the BNL forefront PET program, but the call took a strange turn when Dombo said the congressman wanted to know how to contact Bond over the weekend and asked for Bond's phone number. Puzzled, Bond gave him the phone number of his New Hampshire house where he was spending the weekend, and then waited for a call. No call came.

4 BLOWIN' IN THE WIND

> Politicians are like weather vanes. Our job is to make the wind blow.
>
> —DAVID BROWER

BLINDSIDED AGAIN

Activists make the wind blow by staging protests, getting media attention, and cozying up to politicians.[1] Thanks to their Shoreham boot camp, the activists had this down. The scientists had no such boot camp, and were poorly equipped to provide a counter-wind. The weather vanes began to move.

On Labor Day—Monday, September 1—Forbes and D'Amato sent reporters a "Media Advisory" to expect a "major announcement about Brookhaven," and gave more detail to key reporters so the story would make headlines in the morning papers. The press conference would take place the next day at 11 a.m. in the Press Room of the New York State Supreme Court building in Mineola, a town in the middle of Long Island.[2]

Late on Labor Day, a *Newsday* reporter called Mona Rowe for a comment, alerting her to what Forbes and D'Amato were up to. Politicians about to take actions with significant consequences generally telegraph their intentions ahead of time to leave room for those affected to compromise. This was not happening; Forbes and D'Amato were staging an event for the press that would be impossible to walk back. It was the second time the lab had been blindsided within four months. The first had been the DOE's surprise termination of AUI's management contract on May 1; this one was Forbes's and D'Amato's surprise introduction of legislation to terminate the HFBR.

Rowe's first call was to Bond, who was still away for the weekend. She left a message, the first of many, and began trying to reach other lab officials. Bond returned from his quiet New Hampshire vacation to find his answering machine full of frantic messages.

The next morning, September 2, Liam Pleven's article on Forbes and D'Amato's action and press conference appeared on the front page of *Newsday*'s Long Island section: "Sen. Alfonse D'Amato and Rep. Michael Forbes are planning to announce today that they want the Department of Energy to permanently shut down Brookhaven National Laboratory's main nuclear reactor."[3] Marge Lynch walked to Bond's office, a copy of *Newsday* in hand, and suggested that they should go to the press conference with as many lab officials as they could on short notice. Until this point, lab officials mainly turned their backs on adverse media publicity and press conferences, assuming that the DOE would straighten things out. This was different: the stakes were high and there was nothing to lose.

Bond, Lynch, Rowe, and a few others boarded a van and headed to Mineola, where they ran into other employees who had read the *Newsday* article. These included several representatives of the lab's unions, who saw the action as a threat to the jobs of over two hundred of the HFBR's union workers. One was Gary Zukas, president of the Oil, Chemical & Atomic Workers (OCAW) Union Local 8-431; another was Phil Pizzo, vice president of the International Brotherhood of Electrical Workers Local 2230. The BNL contingent arrived at the Supreme Court building at 11:00 a.m. as the press conference was about to begin, walked through the imposing colonnade entrance to the lobby, and headed left toward the Press Room.

Forbes's aide Diana Weir was standing at the courthouse door. Shocked to see Bond, she dashed back to alert the congressmen. D'Amato picked up the phone and called Martha Krebs, then Weir rushed out to tell Bond that he must call Krebs right away before the conference started. Bond told her that he did not have a phone with him—this was before cellphone universality—nor Krebs's number. Still frantic, Weir ran back again to the congressmen, returned with a cellphone, handed it to Bond, and told him to leave and call Krebs. Bond refused, saying that he wanted to hear what Forbes and D'Amato had to say.

These doings delayed the press conference nearly half an hour. Forbes and D'Amato then took the podium to announce that they were submitting legislation to force the permanent closure of the HFBR. They made the same accusations that they had in the past and continued to repeat: that the plume threatened the off-site groundwater, that the reactor was damaged, that research at the reactor was insignificant, and that the lab had concealed evidence.

"The Lab operates with an air of indifference, with no regard for the health and safety of the people of Suffolk County," announced D'Amato. "We cannot continue to endanger people's health and safety. . . . There isn't a great body of work that has gone on there that is worth jeopardizing the [other] work of the Lab. . . . The High Flux Beam Reactor is in no way essential to the body of research at Brookhaven National Laboratory." Forbes dismissed the reactor as a "damaged and aging" piece of equipment that had had an accident. Both criticized the lab for its "arrogant culture" and "public-be-damned" attitude, and accused the lab of "withholding information."[4] The politicians proudly labeled themselves the "first officials" to call for the HFBR's termination.[5]

At the start of the question and answer period, Bond finally left the conference room to call Krebs, who told him that D'Amato's real fear was that Bond would steal publicity by taking the podium and making a counter-speech. Bond told her that he had come only because otherwise he'd be in his office all day having to talk to reporters one-by-one based on second-hand knowledge of the conference, and that by being present he could give uniform responses to all reporters. Krebs then hung up, having done what she needed: placate D'Amato.

Bond returned to the conference room in time to hear D'Amato boast of being a friend of the lab and trying to strengthen it by closing the reactor—and that he was so supportive of the lab that he had secured funding for RHIC, which he inattentively called a "reactor."

By tipping off the lab, the reporters had transformed the event. Forbes and D'Amato had tried to orchestrate it, but now lab employees were present to be interviewed. The delayed start also gave Rowe and Lynch time to brief reporters and furnish them with provocative questions: Why were the congressmen seeking to abort the DOE's decision process? Why close the reactor when the leak was not a health threat—and, besides, was not from the reactor but from the spent fuel pool? Why had Forbes flip-flopped from his July 14 assurance to lab employees that he knew the reactor was safe? The two politicians found the reception edgier than expected.

Rowe and Lynch were also able to steer reporters toward quotable employees. Asked if the reactor was damaged, Reactor Division head David Rorer said the spent fuel pool was leaking, not the reactor, and that "to close the HFBR now would be like discarding a Rolls Royce that has been pampered by a top-notch crew of mechanics during its entire existence" because of a flat tire. Reacting to D'Amato's remark that the HFBR produced no significant work, physicist John Tranquada said, "Apparently, Senator D'Amato neither reads such major scientific journals such as *Science, Nature* and *Physical Review Letters*, nor talks to anyone who does."[6] OCAW president Zukas said that the politician's action was "motivated by personal political gain," adding that "Senator D'Amato's flippant statement that only 4 percent of the work force at BNL would lose their jobs is far more arrogant than the attitude that he claims the people of BNL have."[7] Bond pointed out Forbes's flip-flop and the fact that the congressmen were short-circuiting a DOE-established process to decide the future of the HFBR, adding that "nearly all of us, our families and our friends live in Suffolk County . . . the events involving tritium emissions at the HFBR have not endangered anyone's health or safety."[8] Another person remarked that calling the tritium plume a "nuclear reactor accident" was like calling a small hole in a car muffler an "automobile accident."

Meanwhile, in Washington, D'Amato's aides submitted to the Rules Committee Bill 1140, the "Long Island Drinking Water Protection Act." Its content was a single sentence: "The Secretary of Energy shall ensure that the High Flux Beam Reactor at Brookhaven National Laboratory is not reactivated."[9]

"Political Cave-In," *Newsday* editorialized.[10] Forbes and D'Amato's action enraged the country's most eminent scientists, such as D. Allan Bromley, the past president of the American Physical Society and long-term Republican who had been Presidential Science Advisor under George

H. W. Bush. He wrote Forbes and D'Amato that their action was "unwise and unwarranted" and would damage US science and technology. Bromley recalled the April statement on the importance of neutron sources by the APS Council, writing that "a strong program in neutron science is necessary if America is to remain a technological leader."[11]

In Washington, Krebs distanced the DOE from the politicians by issuing a statement that "The Department of Energy has not made a decision about whether or not to restart the High Flux Beam Reactor at Brookhaven National Laboratory." The decision-making process, it continued, would include "the views of the people of Long Island, the scientific community and other interested parties," and would be finalized in January 1998, the following year—a hopelessly optimistic deadline.[12]

The next week, Bond was in Washington for an AUI board meeting, and dropped by Forbes's office for their first one-on-one conversation since the press conference. Bond told Forbes that he understood that they may have had different opinions about the HFBR restart, but that he did not appreciate being "sandbagged" by being left uninformed about the press conference. Forbes apologized. Bond then went to D'Amato's office and found D'Amato unavailable. D'Amato's aide Doug Nappi then warned Bond not to attack Forbes when meeting with him in person. Bond kept silent; he'd done just that.

The Friday *What's New* included an entry entitled "Brookhaven: Et Tu Brute?" Park noted wryly, "Most National Labs look to their congressional delegates for support. It works differently on Long Island."[13]

RALLIES: "COME OUT AND EXPLAIN YOURSELF!"

As soon as the Friends of Brookhaven learned of Forbes and D'Amato's action, they began organizing a rally outside Forbes's office in Shirley, a mile south of the lab on William Floyd Parkway.[14]

The FOB members were novices at demonstrations but planned theirs diligently and nerdily. They scheduled it for a 45-minute block of time during lunch, from 11:15 a.m. to noon, so union workers and other staff could get back to the lab without being fined or accused of political activity during working hours. Beforehand they collected and critiqued proposals for signs. Two proposals—"Reason" and "Let the Experts Decide"—were crossed off ("*stinks*," was the judgment).[15] Winners were "Science Fact, Not Science Fiction," "Neutron Science Saves Lives," "Forbes & D'Amato—Over-Reactors," and "We Are the Community!" They printed a one-page handout: "Would the Real Congressman Forbes Please Stand Up?" contrasting the congressman's July 14 assurance to lab employees with his recent statements. They circulated a petition demanding that Forbes and D'Amato "postpone introducing any legislation which short-circuits" the DOE's process for making a decision about the reactor.[16] They sent a press advisory alerting the local media to an unusual event: a showdown between scientists and their own congressman.

Two days after the Forbes-D'Amato press conference, on the morning of September 4, five hundred people showed up in the parking lot of the building where Forbes's office was located, a drab concrete structure now used for outpatient surgeries, and a ten-minute drive from the lab.[17] Joanna Fowler carried bundles of flyers in her red truck; others brought a bullhorn and an American flag. Shanklin carried the petition with fifteen hundred signatures. The parking lot was full, and some demonstrators had to pull over on the side of the parkway and walk several minutes.

Everyone nervously milled around the parking lot for a few minutes, unsure of how to start a demonstration. "We felt like geeks," Shanklin said. Eyes turned to him. He climbed on the back of Fowler's truck, grabbed the bullhorn, draped the US flag by his side, looked to the window of Forbes's second-floor office, and shouted, "You've made these statements about the lab that are not accurate!" Not the usual heated rhetoric one hears at demonstrations, but it started to galvanize the assembly, whose mood rapidly turned from awkward to upbeat. Energized, Shanklin continued, "Come out and explain yourself if you can!"

Forbes did not appear—he was in Washington—and the now worked-up crowd began hooting and jeering. "It appears that our congressman can't justify his position," Shanklin continued, drawing cheers.

Other speakers followed. Harold Atkins, a physician from Stony Brook University's hospital, expressed outrage at having to stop his research at the

Figure 4.1
Brookhaven employees' first rally in front of Forbes's office.

HFBR on radionuclides that reduced pain from cancer. Zukas said, "The senator and congressman are obviously misinformed of any actual or perceived hazard presented by the HFBR to the drinking water of Long Island." Rich Sanniola, a union official from the Safeguards & Security Division, who was vice president of Local 37 of the Long Island Guards Union and who worked regularly at the lab gate, pointed out that the pool of protestors who regularly appeared at the gate billing themselves as representing "the community" typically numbered twelve people.[18]

Union official Phil Pizzo demanded to know why Congressman Forbes wanted to send jobs off Long Island by shutting the reactor. Madeline Windsor, of the Technical Information Division, said that though a Republican, she was tired of "irresponsible politicians." Alfredo Luccio, an Italian immigrant who worked at the AGS, said he had come to America "because of the fairness and justice of this country" only to learn from the actions of Forbes and D'Amato that "there is no fairness, no justice." Physicist John Tranquada said that because he was trained "to base my arguments on the facts" he could only admire Forbes's "chutzpah." The lesson to local high-tech businesses, he added, was that if they used potentially hazardous chemicals, even safely, Forbes and D'Amato could "shut them down permanently without discussion."[19]

Most dramatic was Gar Harbottle of the Chemistry Department. Harbottle's big, disheveled Einstein-like hair and unique posture—he stooped with his left hip out to one side, Marx Brothers–style, crooking his left arm behind him to rest on that hip—made him look like a caricature of a mad scientist. His research program at the HFBR was readily explainable: he used neutrons to date and discover the origin of ancient artifacts, allowing scientists to map early trade routes. Harbottle's words were informed, but they were spoken with the passion and anger of the activists; he held up and shook a carton of orange juice and shouted, "This contains more intense radiation than the plume!" though exactly what kind of measurement he had in mind was unclear. Patches of black sand on Long Island beaches, he continued, contain naturally occurring radioactive thorium. "There is 'serious radiation' on the beaches of Long Island," he concluded, "Forbes shouldn't close the HFBR; he should close Westhampton Beach!"[20]

Forbes had been in Washington the day of the rally meeting with D'Amato and Peña. Peña told the two congressmen that he would await the results of the Environmental Impact Statement before deciding whether to restart the HFBR. Peña wanted to keep the meeting private so as not to embarrass the politicians, but wound up having to make a press statement.[21] Krebs called Bond afterward, telling him that Forbes and D'Amato did not understand the Environmental Impact Statement process, and emphasized that Bond should support the secretary's plan.[22]

When Forbes returned that weekend, he held a series of town meetings, declaring that he would not change his mind under "any condition." Members of FOB participated at these meetings. When Forbes accused the lab of organizing a demonstration against him when its employees were getting paid with federal dollars, FOB members were able to counter by pointing out that it was during lunchtime. Asked to explain his July 14 remarks assuring Brookhaven employees that he knew the HFBR was safe and should be restarted, Forbes mentioned "revelations" that included the discovery of a "new plume" of tritium. Lab employees said that what he called revelations were discoveries of the historical survey into previously allowable practices, and that what Forbes called a "new plume" was not a plume but the spike of unknown origin or a one-time event due to the discharge of condensate during a change-over in the air filtering system, which had elevated the tritium concentration at the sewage treatment plant somewhat but was still below the drinking water standard.[23] But that episode had mandated a press release, bringing it to the attention of activists and politicians, who naturally used it as another data point illustrating the lab's menace.[24]

At a Forbes town meeting that weekend, a cancer victim charged the congressman with "condemning people to die and die painful deaths by closing that reactor." When another, Lois Mitchell of Poquott, spoke on behalf of the cancer victims who could be denied treatment if the congressman succeeded in closing the reactor, Forbes responded that she was being "unfair" by bringing up a hypothetical possibility. "Who's being unfair?" she retorted. "It is not 'unfair' to rely on the experts who know what is going on instead of politicians." At another town meeting, a scientist in Brookhaven's Biology Department bluntly told Forbes: "Things you say aren't quite true, and you have to stop saying them."[25]

On September 8, Forbes sent a four-page letter to constituents stating that his goal was to shut down the "leaking" HFBR. Once again, in contrast to his July statement, he wrote, "It would be foolhardy and quite irresponsible to knowingly permit the re-start of this outdated, aging nuclear reactor with such a questionable history," and later, "You can always count on me to be honest with you."[26] The letter contained an "official constituent opinion survey" asking constituents to vote by checking one of two boxes. The first was labeled "YES Mike, you are correct in forcing the permanent shut down of the leaking HFBR nuclear reactor at Brookhaven Lab. The problems there warrant a tough decision on your part and, it is the right approach to years of environmental mismanagement." The other was "NO Mike, this controversy about Brookhaven Lab is overblown. You should allow the nuclear reactor to reopen."

Several lab employees, including Jennifer O'Connor, noticed that many women, and no men, had received the letter; in households with couples registered with the same political party the wife received Forbes's letter and the husband had not.[27] The women concluded that Forbes had mailed it only to female constituents, assuming that they would be most easily swayed by the rhetoric. Outraged, Brookhaven Women in Science sponsored a lunchtime workshop, "How to Write to Elected Officials," a few days later. "All are invited; bring your lunch, writing paper and sharpened pencils."[28]

* * *

Conflicts were also breaking out between activists and lab employees, sometimes spinning out of control. Union President Zukas called Bill Smith, and the two wound up threatening each other. On September 11, Zukas was arrested and charged with aggravated harassment after Bill Smith had accused him of making a death threat.[29] Zukas was released and ordered to respond before the Shelter Island Justice Court. Smith issued a press release about this, and *Newsday* carried the story.[30] Smith had had his own scrapes recently; on August 30, he had been arrested for failure to yield right of way to an emergency vehicle and second-degree obstruction of government administration.[31] Two and a half weeks later, on September 29, Zukas appeared before the Shelter Island Justice Court. The case ended in an "adjournment in contemplation of dismissal," and Zukas was freed with instructions to keep away from Smith for six months. That same day, four anti-HFBR activists demonstrated at DOE headquarters in Washington, demanding the reactor's permanent shutdown. After blocking the building entrance, they were arrested, ticketed, and released.

* * *

Forbes had been unimpressed when he learned of the rally outside his office, and confidently told a lobbyist that that would be the end of it. They're scientists, Forbes said; they've never demonstrated before and never will again.

"The minute we heard that," Shanklin said, "we started planning the second rally." He told other FOB members, "Here's the plan: Bigger truck, bigger flag, bigger crowd." On September 19, lab employees staged another lunchtime rally at Forbes's office. Someone indeed found a bigger truck. Marotta, an FOB member who was also a volunteer firefighter and the lab's fire department manager, located a garrison flag—20' × 38', the largest flag used by the US

Army and only on special occasions—and mounted it behind a speaker stand. Instead of a bullhorn, they had a sound system. This time the rally attracted 1,200 people, who had to park even farther down William Floyd Parkway.

Several women, including O'Connor, denounced Forbes's mailing as "political junk mail" and said they were appalled that it seemed to have been targeted to female constituents. O'Conner told the crowd that she had worked inside the reactor building during two healthy pregnancies, and continued, "Not only was this poll a waste of taxpayers' hard-earned money, but it was also an insult to women. How dare [Forbes] assume that women would be more ignorant and sympathetic to his cause. Does he think that we're all a just a bunch of Peggy Bundys? [a TV character who was the symbol of a helpless housewife]. My message to Congressman Forbes is that women make informed decisions."

Union representatives spoke of the anticipated job losses of the actions, while others ridiculed Forbes's comments about the lab's supposed threat to health, and pointed out the value of the HFBR to researching diseases like cancer. Catching the rhetorical spirit, Jeff Coderre of the lab's Medical Department said: "But how long will this continue? The extremists [are already thinking of going] after the Medical Reactor. Next will come the AGS, RHIC, and then the whole Lab. [So, Forbes's and D'Amato's attempt to close] the HFBR is a dangerous precedent with far-reaching implications for the future of scientific and medical research across the country."[32]

This time Forbes's employees called the police, reporting that demonstrators were blocking their parking lot. Shanklin intercepted them and pointed out that the rally was peaceful. Looking at his watch, he assured the police officers that the parking lot would be completely empty of demonstrators in twenty minutes—for most had to be on their shifts in half an hour—but, Shanklin added, the demonstrators would be thrilled if the police broke up the rally by grabbing and arresting people in front of TV cameras. The officer paused, and said, "You know, this might be a good time for us to take a smoke break."[33]

STAR, and the militant activists, attacked the protest. "I'm appalled," Bill Smith said, "that the employees of BNL are attempting to justify their legacy of pollution of Long Islanders with taxpayer dollars," without indicating exactly how the money was supporting these lunchtime protests. Karl Grossman warned that the scientists were violating the 1939 Hatch Act prohibiting "pernicious political activity" by federal employees—evidently not realizing that they were not federal but AUI employees—and called them "in total denial of the environmental mess at the lab."[34]

REHASHINGS

After Bond got a copy of Forbes's letter to constituents he wrote a long memo to lab employees itemizing and correcting statements that Forbes was making in letters, to reporters, and at town meetings.[35] Forbes had written that the HFBR was leaking; it was not, Bond wrote, for the leak was in the fuel storage pool, which "has nothing to do with the way the HFBR operates." Forbes wrote that nuclear-powered US Navy ships were routinely taken out of commission at about the age of the HFBR; Bond noted that this was due to hull fatigue and new technology—and that the nuclear-powered USS *Enterprise*, powered by six reactors and built before the HFBR, was still in action and projected to last at least another fifteen years before it, too, would succumb to hull fatigue. Forbes had referred to seven plumes at BNL; Bond pointed out that only one of the plumes that Forbes had mentioned was related to the HFBR—and that not to the reactor but the spent fuel pool—and that the lab had been addressing the other six, all but one chemical-related, for years.

Forbes had also claimed that the lab had never investigated the 1986 well with tritium, that it had become a Superfund site because of horrible practices, that the low priority that the lab gave groundwater monitoring was a sign that it was uninterested in the environment, and that the lab had hidden the tritium leak the previous October. Bond responded: the 1986 well had been closed for chemical contamination and its tritium level was far below the standard; the lab had become a Superfund site because of routine, if now inappropriate, practices that mostly met regulations at the time, practices shared by farmers, dry cleaners, and gas stations; an extensive program of groundwater monitoring had been done for years with the installation of a network of monitoring wells but these had been to assure that contaminants were not carried off-site by groundwater; and it was not October but December that the first hints of tritium were discovered but they still had to be confirmed.

One AUI trustee suggested that Bond send his employee memo to Forbes's constituents. Bond thought this unwise, for it would appear that the lab was engaging in politics. But AUI did print a condensed version of Bond's memo as a full-page ad in *Newsday*.[36] The ad consisted of two columns. The left-hand column listed "What Congressman Forbes Says," and the right "The Facts." "[T]he *truth* is at stake," the ad declared, and a democracy depends on citizens knowing the truth to make wise decisions and protect itself.

AUI, a departing organization fired by the DOE, might have been expected to want to wash its hands of Long Island in disgust; nevertheless, it took out the full-page ad to lay out well-known facts.

The ad did not deter D'Amato and the activists, who continued to repeat the claims listed in the left-hand side of the ad as if there were nothing on the right-hand side.[37] The two congressmen called, again, for extensive environmental reviews of the lab, as if the already-completed reviews by the DOE Office of Oversight and by the CDC's sister agency the Agency for Toxic Substances and Disease Registry (ATSDR), as well as ongoing reviews by the Department of Energy, the Environmental Protection Agency, Suffolk County, and New York State—each an organization that citizens had every right to look to for the most objective analysis—did not exist or were incompetent.

Bond continued to spar with Forbes on radio interviews, and TV programs. One was a 10-minute call-in morning show on the local Channel 12 in which a reporter asked Bond and Forbes about the reactor. Each rehashed what they had been saying for weeks—Forbes that the reactor should be shut because it was dangerous, Bond that it was safe and that the leak was from its spent fuel pool, and that the DOE's process involving an EIS should go ahead as planned. The exchange had enough point–counterpoint that the station set up a half-hour television debate between the two, broadcast on September 14, replaying the same issues in more detail. At one point the moderator said that Bond was "in no way" associated with AUI, and was perplexed when Bond corrected her. In her final question the moderator

A Public Response to Congressman Forbes
From Associated Universities, Inc.

To Our Friends and Neighbors on Long Island -

Recently, Congressman Michael Forbes sent a four-page letter to many, but not all, of the residents of the First Congressional District. In that letter, he continued his call for the closure of the High Flux Beam Reactor (HFBR), a small nuclear research reactor at the Brookhaven National Laboratory which is used exclusively for scientific experiments.

What has particularly upset us at Associated Universities is that this mass mailing, paid for at taxpayer Expense, makes serious misstatements - presented as if they were facts.

Associated Universities has managed Brookhaven for the federal government throughout the Lab's illustrious 50-year history. During that time BNL has consistently produced world-class research, much of it done at the same reactor Congressman Forbes now wants to close!

We at Associated Universities have already announced that we will not be in the running to manage the laboratory when a new management contractor is selected in the next few months, so we have no organizational stake in the issue over whether or not to continue using the HFBR. But we do believe the people of Long Island need to have the correct information before they make up their minds on this issue. That's why Associated Universities is using its own funds to publish this ad - to clarify and correct some of the issues raised in Congressman Forbes' letter.

What Is At Stake Here?

You might be asking yourself, "If AUI has nothing at stake, why take out this ad?" That's a good question, and, quite simply, we feel that there are three important things at stake here:
• Thousands of valuable jobs, especially for scientists and engineers, are in jeopardy.
• Important national scientific research programs are at risk.
• Most importantly - the *truth* is at stake, and truth is something every citizen deserves when being asked to make a decision of this magnitude.
• What follows is a comparison between what Congressman Forbes says, and what actually happened.

What Congressman Forbes Says:

• "This note is necessary because of the independent decision I have made to force the *permanent* shutdown of the leaking HFBR reactor at Brookhaven National Laboratory."
• "[BNL] became a designated 'Superfund' site in 1989 because of the errant waste disposal practices it followed for much of its 50-year history and thus became eligible for priority clean-up."

• "In 1992 after concerns were raised about the slow pace of clean-up at BNL, an agreement was reached between DOE, the EPA, and the State Department of Environmental Conservation to ensure proper compliance at BNL with Superfund requirements...At that time, Lab officials acknowledged the need for extensive groundwater monitoring but assigned it a very low priority."
• "Last October, a routine sampling of wells near the nuclear reactor uncovered extraordinary high levels of tritium, one measuring 32 times the drinking water standard. That's bad enough but when Lab authorities failed to notify the public until this past January, it precipitated the crises that has gripped the Lab to date."
• Finally, at our urging the [EPA] launched a comprehensive investigation at BNL this past May that found numerous violations of environmental rules and regulations...They uncovered the existence of a pipe discharging into a nearby wetland from the on-site hazardous waste management building where mixed wastes, solvents, acids and other toxins are stores! They were cited for exceeding by more than 50 percent the mandatory Clean Air safety standards in steam that results from the operation of a boiler used to burn waste."

The Facts:

• *The reactor is not leaking.* The tritium plume came from a leak in the pool used to store spent reactor fuel. This storage pool has nothing to do with the way the HFBR operates. The HFBR could have continued to operate safely.
• *Waste-disposal practices followed at BNL and which led to BNL's being named a Superfund site in 1989 were all in accord with the regulations of the time.* This clean-up is evidence that management has been following increasingly more enlightened regulations, not violating them.
• That 1992 agreement was the culmination of a three year period of negotiation and public involvement, *a process that is required by law* after a site is named as a Superfund site. In 1992, BNL extended its network of monitoring wells, giving highest priority to its southern borders, to determine if contaminants were going off site. Chemical plumes were found, but they were too deep to affect residential wells.
• *BNL did not delay notification.* The first indication that tritium was in the groundwater came last December, *not* October, when results from monitoring wells showed very low levels of tritium. Since this result was the first indication of a problem, it had to be confirmed with another test, so confirmation of the high level came o January 10.
• None of the violations in the EPA's compliance audit of BNL pose a threat to workers, the public or the environment. The pipe Congressman Forbes refers to is a stormwater drain under a road, no a discharge pipe from a hazardous waste storage, and the area it drains to is not a wetland. Finally, the boiler emissions Congressman Forbes cites came while a new boiler was being tested to see if it met our standards. It didn't, and we rejected it without prompting by EPA or anyone else.

For a more complete discussion of these issues, please visit our web site at "www.aui.edu"

AUI
AN ACADEMIC PARTNERSHIP · DEDICATED TO SCIENTIFIC RESEARCH.

Figure 4.2

AUI's "truth is at stake" advertisement. Courtesy Associated Universities, Inc.

asked Bond if the lab would do whatever the DOE recommended regarding restarting the reactor. "We don't have much choice," he said.[38]

Bond's motive in continuing to respond to Forbes was not only to correct statements but to support the morale of lab employees, many of whom were having a difficult time coping; visits to the mental health services at the lab had shot up that summer. Even certain experienced and conflict-hardened administrators were personally devastated by the repeated brutal attacks on their integrity, and asked Bond to replace them. These included Sue Davis, who was exhausted by the vituperation that antinuclear activists had repeatedly directed against her; Bond appointed Bill Gunther to replace her.

The various civic groups were now jockeying for position over who counted as speaking for "the community." The issue was important, for it had two consequences: first, those who can convincingly claim to speak for the community gain a special authority, especially in a democracy; second, it suggests that all other groups are mere special interest groups and do not have a legitimate standing in the discussion. STAR, in particular, was aggressive in claiming to be the voice of "the community." But so did the Long Island Association, a prominent local business group, who urged that Peña allow the EIS process to be completed.[39] A coalition of local community groups held a press conference near the laboratory's main gate at which they criticized Forbes and D'Amato's attempts to shut the HFBR. "What's most upsetting to the community around the Lab," said Connie Kepert, head of the Longwood Alliance and former president of Affiliated Brookhaven Civic Organizations (ABCO), an umbrella group for over 40 Town of Brookhaven civic groups, "is that Forbes and D'Amato have circumvented the process in which many community groups have been participating. We are insulted and upset that they decided to preempt that process and make a decision before the ABCO, the Long Island Neighborhood Network, the Long Island Progressive Coalition (LIPC), individual local civic groups and the Community Work Group, which has monitored BNL issues" were consulted. "The Community Work Group [CWG] has been monitoring the Lab for two years," added Judy Pannullo, executive director of LIPC and a CWG member. "We're horrified that Forbes and D'Amato never came to us to discuss this problem. It seems that these politicians are reacting to the East

End environmentalists who are making the most noise. . . . We have to put democracy back into the process."[40]

* * *

October 1 was the eve of Rosh Hashanah, and Rabbi Joseph Topek was presiding over the ceremony at Stony Brook University.[41] Rabbi Topek spoke of Tashlich, the Rosh Hashanah ritual in which, as per Micah 7:19, one throws one's sins into the pure flowing waters. Topek made a flippant remark that, thanks to the lab down the street, he was not sure about the purity of the local waters. Michael Marx, a Stony Brook physicist who performed experiments at the lab, was appalled, and after the ceremony upbraided his friend Topek and explained that the tritium at Brookhaven was not in the local waters.

Topek's remark was flippant, but that's what troubled Marx. The activists had successfully popularized the charge that Brookhaven was a polluter. Thanks to the continuing publicity the activists had generated, "everybody knew" that the lab polluted the groundwater and that any groundwater pollution anywhere on Long Island must be the lab's fault. It had become a meme.

Marx was able to turn around Topek's perception because of Topek's personal friendship with Marx, his respect for Marx's expertise, and his trust in the EPA and other studies that Marx showed him. Not everyone shared this trust, and the activists sought to erode trust in any expert or institution that challenged their claims and spread that distrust as widely as possible. Even in the corridors of Stony Brook University, stickers appeared on some doors with a profile of the HFBR over which was superimposed the "NO" symbol—a red circle with a slash through it. When one Philosophy Department graduate student was challenged about the judgment expressed by the sticker, and shown factual studies, the student was perplexed: "Well, why don't the scientists speak up about it, and make these facts known?"

RHIC, AGS, NSLS

In the fall of 1997, in the huge ring tunnel to the north, Brookhaven's scientists and engineers continued to piece together elements of the Relativistic Heavy Ion Collider (RHIC), aiming to complete and operate the machine in 2000. The single biggest piece to arrive was the time projection chamber (TPC), a six-ton, $10 million "camera" nearly 14 feet long and 14 feet in diameter. The TPC would collect tens of millions of bytes of data about each collision and use the information to "project" the collision back in time to give the scientists a picture of how it had unfolded. Such collisions would yield information about "quark soup" and other forms of matter that have not existed since the time of the Big Bang, and would shed light on the formation of the Universe.

Meanwhile, the RIKEN-BNL Research Center co-sponsored by Japan and the United States—whose founding document was signed April 30, the day before Peña terminated AUI's contract—opened on September 22, attended by dignitaries from the United States and Japan. RIKEN had been established to foster young physicists in both experimental and theoretical research at RHIC.

Experimenters at the AGS, which was soon to become the injector accelerator for RHIC, made two significant discoveries in high-energy physics. One was of a rare type of subatomic particle called an "exotic meson." Three weeks later, another AGS team, E787, announced the discovery of a rare type of particle decay that happens only once or twice in every 10 billion decays of particles known as kaons.[42] Yet a third key experiment taking shape at the AGS, E821, was called "Muon g-2," the highly sensitive measure of the difference from the theoretical prediction for the way that the particles "wobble" while circling in a ring with a constant magnetic field, which indicates the overall adequacy of the current state of quantum electrodynamics.

The National Synchrotron Light Source (NSLS) had a close call that summer. Light sources were increasing in importance, thanks to their critical role in materials science, medicine, and biology, but the cost of their operation was skyrocketing and taking up a quarter of the DOE Basic Energy Sciences budget. Worried, Krebs created a committee in April 1997 to visit and evaluate the four US synchrotron light sources to see if they were all needed and what the consequences would be if the DOE shut down one or more of them.[43] Rumors circulated that the specific target the DOE had in mind was the NSLS, the oldest and most antiquated of the four. The committee, headed by Robert Birgeneau, visited Brookhaven on June 25–26.

The committee released its report in early October, and it was counter to the DOE's budgetary hopes. The "outstanding performance" of the NSLS, the report stated, made it a national resource and one that should be "adequately funded, upgraded and modernized in a timely fashion to serve better the national needs."[44] "The Committee concludes unanimously that shutdown of any one of the four DOE/BES synchrotron light sources over the next decade would do significant harm to the nation's science research programs and would weaken our international competitive position in this field."[45] The report even recommended that the NSLS receive $36 million more in upgrades.

ANOTHER STINGING JOLT

On September 10, the comprehensive Facilities Review, initiated in April to seek out historical activities that might have led to environmental issues, had pinpointed twenty-one significant environmental findings, twenty of which were already known and fourteen of which were already being corrected. The previously unknown finding was an accidental release forty-five years before, in 1972, of about 100 gallons of oil to a drain in the AGS building.[46] And yet, five days after the encouraging review came out, another more serious issue was the discovery of 60,000 gallons of radiation-containing water in the exhaust air duct of the now-defunct BGRR. Though the BGRR had closed in the mid-1960s, the AEC and its successor agencies the Energy Research and Development Administration (ERDA) and DOE had never provided money to fully decommission it. In the late 1990s, Congress began appropriating special funds for cleanups, but Brookhaven had never received these funds as it was such a low priority site compared with areas with truly serious cleanup issues such as Hanford and Rocky Flats. In a decade and a half of operation there had been occasional fuel failures at the BGRR, so the ducts were known to have significant radioactive contamination but had never been closely examined; finding water in them was a surprise and indicated a significant oversight in monitoring on the part of both BNL and the DOE.

The analysis of water samples came back on October 3 and indicated high levels of Cs-137 and Sr-90, as well as small amounts of other isotopes including transuranics. Scientists put in a call to R. W. Powell, a retired employee who had worked on the BGRR from the beginning; Powell said that periodic investigations had come across occasional water but nothing to the extent of the current finding.[47] Investigators went to extraordinary lengths to make sure to safely remove the water, even going so far as considering the faintest of possibilities— that the transuranics could reach a critical mass. A plan to remove and dispose of the water was drawn up and sent to the DOE.

O'TOOLE DEPARTS

Tara O'Toole abruptly announced her resignation on September 18, effective October 12. The DOE's press release said nothing about the reasons, only that she planned to "rest, reflect, and write."[48] Why she resigned just then was a mystery. She had taken upon herself the goal of cracking down on environmental issues at the DOE labs—with such determination that she had acquired the nickname "Terror O'Toole" even within local DOE offices—but had not yet achieved it. Furthermore, she had been largely responsible for the firing of AUI, and her announcement occurred just four days before a key milestone in the bidding process. For some unclear reason, she was walking away from the waves that she had so effectively created.

O'Toole appeared defensive when questioned by an *Energy Daily* reporter. She had come down hard on the national labs to protect their "very existence," she said. "It's not big, bad DOE doing this to [the labs]. It is DOE, the political acumen in this building, trying to save them."[49]

After reading this, Bond wrote O'Toole, "Surely there must have been another way to wake up the DOE complex than to fire AUI without warning and put BNL in a position of chaos for months on end." The attacks by elected officials on the lab, he wrote, the "incredibly rapid pace" to find a new contractor, and all the changes to follow "are not the ideal way to make improvements." The lab and AUI were making many significant improvements in April, but the May 1 termination of the contract took a lot of steam out of them.[50]

"[F]iring AUI was necessary," O'Toole shot back. "I hope it was a big enough shock to people in the DOE system to provoke the changes that are so difficult to accomplish . . . and so necessary if the missions of DOE are to survive, let alone prosper. . . . I do not regret any of the BNL decisions, which I realize seem partly inscrutable to you, sitting on Long Island, far from Washington DC. One of the perils that confronts democracy is the distance, figuratively and literally, that separates citizens from those who try to do the work of governing."[51]

"I REALLY don't understand much of politics and I'm not sure I want to," Bond answered, for "facts seem to be irrelevant much of the time."[52]

"Peter—there are no 'facts'—only data that need to be interpreted," O'Toole replied.[53] Nearly twenty-five years later, O'Toole remained firm. When asked if it had been necessary to fire AUI, she replied, "Yes. If we didn't do something dramatic, others would have. . . . There was real pressure to close labs. . . . You had a political problem that the scientists for the most part were oblivious to. . . . In the background was the waning of the post–World War II American embrace of science and technology. . . . The scientists kept trying to talk about rems and risk to health and they didn't get that they weren't trusted."[54]

5 COMPETITION FOR THE NEW CONTRACT

[A] new way of doing business at the Department of Energy.

—FEDERICO PEÑA

CONTENDERS

In parallel with all these events at the laboratory, an entirely different high-stake drama—equally confused and chaotic—was unfolding in the search for a new contractor.

Peña clearly expected a spirited competition. Surely, the chance to bid on a $400 million per year contract with his agency would attract top-flight contestants. The lure was not the victor's $7 million maximum annual fee but the perks.[1] A university would get prestige, an influence over research, and joint faculty appointments; a corporation could profit from technology transfer programs, recruit skilled employees, and work closely with the DOE.

Competing for a DOE contract was normally a stressful process lasting a year and a half, but Peña intensified the ordeal by shortening it to six months, promising to have a new manager in place by November. That changed everything. To assemble a team, develop a structure, and write a proposal was effectively impossible for all but large corporations with existing DOE contracts. Peña's requirement, and additional ones in the Request for Proposals (RFP), meant that the days of AUI-like groups of universities managing national labs whose primary aim was promoting science were over, at least as stand-alone entities.

Peña chose Steven Silbergleid, the chief counsel for the DOE's Chicago office, to oversee the competition.[2] Careful and efficient, Silbergleid was canny enough to sense danger, and accepted only on condition that if he had any urgent questions Peña would respond within a day. Silbergleid then brutally hacked away at the standard contracting process, curtailing the usual internal discussions and omitting routine steps. His most draconian action was requiring those with key roles in each bid—such as proposed lab directors—to sign a commitment to taking the position before submission, a step that cut out often-lengthy negotiations. Silbergleid said, "We were basically telling people, 'Take it or leave it.'"[3]

Peña had promised that he would seek Long Island community input. It was a noble idea, but Silbergleid, who spent much of May sifting through the comments, was appalled. "I don't mind taking comments at a public meeting," he said. "But that's very different from allowing the public to make written recommendations for what you think is important to consider in selecting a contractor. Some of the comments were on the outrageous side. You were expected to answer them all, and you couldn't say, 'This is so dumb I'm not going to consider it.'"[4]

The competition kicked off May 28 with a notice in the *Commerce Business Daily*, followed by guidelines, a workshop, and a lab tour that attracted seventy-seven potential bidders from fifty-two institutions and included representatives of large corporations, several universities, and many small companies that saw themselves as potential subcontractors. It was a diverse and competitive bunch. The requirements of the formal, six-hundred-page RFP, however, issued on July 18, scared off nearly everyone, for the document seemed to make two partners necessary, one to handle operations—budget, administration, and ES&H—and another to supervise research. Many large corporations lost interest, furthermore, once it became obvious that they would not make significant money from radiological cleanup. While environmental restoration at Hanford, Savannah River, and others involved billions of dollars, at Brookhaven the tritium was harmless and already being remediated, and the small amounts of other radiological isotopes were minor, staying in place, and also being cleaned up. Universities, meanwhile, were apprehensive about the unfamiliar, time-consuming, and risky path of having to forge managerial relationships with corporations within weeks, and very likely to have to play junior partner.

For a while AUI planned to stay and fight, convinced that its half-century of experience managing a world-class national laboratory would be a tremendous asset for any team. After two months of rejections by potential partners, and a series of negative signals from the DOE, the trustees gave up, making their action public on August 5.[5] Schwartz, who for three months had been seething with anger, composed a bitter memo outlining AUI's outstanding record managing the lab and detailing the unjust actions that had destroyed AUI's chances:

> Three months ago, on May 1, 1997, Federico Peña, the Secretary of Energy, terminated AUI's contract "in the best interests" of the government. . . . Conditions at Brookhaven did not and do not endanger anyone's health or safety . . . the termination action by the Department was a totally dispropor-tionate response to the problems, particularly in view of the fact over the last 10 years DOE regularly gave AUI official evaluations of "Good" to "Excellent" in all categories and consistently graded AUI "Excellent" in overall manage-ment . . . the Department gave AUI no opportunity to discuss the initiatives and corrective actions that were underway before the abrupt termination. . . . AUI could have brought the strength of 50 years of successful management of science to a management team . . . [but thanks to DOE actions] a proposal containing AUI was doomed to failure.[6]

The militant activists lost no time belittling Schwartz. The *New York Times* quoted Bill Smith as saying that Schwartz was behaving like "a spoiled child."[7]

By early August, two weeks after the RFP, only two teams were in the works, one a partnership of Battelle Memorial Institute and Stony Brook University, the other of Rochester Polytechnic Institute and the Westing-house Corporation. With Peña's compressed bidding process the final pro-posals were due in three weeks, on August 28.

The Battelle Memorial Institute was an international science and tech-nology research corporation based in Columbus, Ohio. Founded in 1923, it worked closely with the government and military on nuclear projects, pro-viding fuel rods for the first nuclear-powered submarine, the *Nautilus*.[8] The company had a huge incentive for seeking to manage Brookhaven. The state of Ohio had sued the company for $70 million in back taxes on the grounds that it was for-profit rather than not-for-profit.[9] Though Battelle managed the Pacific Northwest National Laboratory (PNNL), that counted for little, given that research provided only a small fraction of PNNL's budget; managing Brookhaven would firmly stamp Battelle as nonprofit.

Battelle representatives visited Brookhaven to introduce themselves. "We expected a talk about their research but they kept harping on how strongly they were committed to nonprofit activities," Rowe recalled. "You got the sense they were protesting too much."[10] That evening, when PNNL director and Battelle employee Bill Madia took fellow lab director Samios out to dinner, Samios told Madia that Battelle would need an academic institution on the team, and they ran down a list of possibilities. Samios pushed Columbia University, his alma mater, and asked Madia to contact Columbia professor T. D. Lee, whose Nobel had been awarded for work at Brookhaven. But Lee was leaving for a two-week trip to China—and at the competition's break-neck pace this all but closed off the chance to partner with Columbia. Samios also suggested what was then called the State University of New York at Stony Brook, and now Stony Brook University, whose president was Shirley Kenny.

Kenny had been Stony Brook's president for three years, having succeeded John H. ("Jack") Marburger III. She spoke slowly, in a heavy Texas accent, in a way that disguised her energy and ambition. Rank-and-file faculty regarded her as aloof, but she was politically astute and had the determination that an ambitious upcoming university needs; Battelle's Madia was shocked to discover that she was as combative and effective as he was. Kenny also knew how to give good press conferences, enlivening them with deadpan humor. When she made her first speech after arriving at Stony Brook, she got laughs with the (accurate) remark that she was pleased to be the first president of the University who was not a white male physicist named John. Stony Brook was an obvious contender in the competition as the university closest to the lab. Established in 1957, too late to be a founding member of AUI (though by 1997 it had a member on the board), it had built strong science and technology departments, but had not achieved the status of other major research universities, and Kenny was aggressively look-ing for ways to elevate its profile. She had a specific goal: to get Stony Brook in the Association of American Universities (AAU), a prestigious group of 63 research universities, to which one had to be voted in by three-quarters of its members.

On the surface, Rensselaer Polytechnic Institute (RPI), a small but well-respected engineering college in Troy, New York, seemed an unlikely bidder. Its Physics Department was strong not in particle but in applied physics, materials science, and astronomy, areas that did not align well with Brookhaven's research. But RPI's dean of engineering was Richard Lahey, an internationally known specialist in nuclear physics and engineering, as well as nuclear safety technology. He had an MA in engineering mechanics from Columbia and a PhD in mechanical engineering from Stanford, had been at Rensselaer since 1975, and its dean of engineering since 1994. Lahey was difficult to overlook, literally. He stood six and a half feet tall, was a former Navy Seal, and was outspoken and aggressive. Aside from his expertise in nuclear engineering, he was also valued for his extensive knowledge of wine. Lahey knew Brookhaven, having been on visiting committees, and was able to reassure lab employees who worried about the fact that his interests lay in engineering rather than high-energy physics. Lahey saw no problems with an RPI role in managing Brookhaven. In his eyes, the lab's principal problems involved nuclear engineering, and he saw great possibilities for expanding RPI's own science and engineering programs. Lahey found RPI's acting provost, physicist Jack Wilson, enthusiastic. Lahey had more dif-ficulty with RPI President R. Byron Pipes. Under financial pressure, Pipes was seeking to downsize RPI and was wary of new projects—which made him unpopular with the RPI faculty. Pipes was also prone to vacillate, and was lukewarm about bidding for Brookhaven; there would be little money in it and possibly significant liability. Nevertheless, Pipes encouraged Lahey while reserving the right to make a final decision, and Lahey proceeded to contact counterparts at Columbia, MIT, Stony Brook, and Westinghouse.

The Westinghouse Electric Corporation, based outside Pittsburgh, was well positioned to bid for Brookhaven. The company had built an advanced particle accelerator in 1937, had military and other government contracts, supplied the world's first commercial pressurized water reactor in 1957, and had built and operated its own research reactor. Westinghouse also had experience with groundwater contaminated by radiation, with monitoring test wells, and with dealing with environmental concerns of surrounding communities. By the 1990s Westinghouse, through subsidiaries, was involved with numerous environmental restoration contracts for the DOE on the order of billions of dollars, including the Savannah River facility, and, starting in 1995, a lucrative subcontract to handle nuclear wastes from a site in West Valley, New York, a contract due to be recompeted in 1999.[11] In the DOE's eyes, as well as those of New York State congressmen, Westinghouse was a front-runner. In June, a Westinghouse corporate jet took a team on a tour of universities to hunt for partners for the BNL contract, and received the best welcome after it docked at Troy on June 23. Lahey had worked with Westinghouse in the past, Provost Jack Wilson had grown up in Pittsburgh, and both were thrilled to partner with Westinghouse.

SCRAMBLING TO MEET THE AUGUST 28 DEADLINE

The Stony Brook–Battelle partnership, which began to take shape in mid-July, got off to a rocky start. Battelle realized that the DOE's overriding concern was to have a team able to manage ES&H effectively, knew that the DOE knew it, and viewed itself as the prime contractor, envisioning Stony Brook as having the lesser role of managing the lab's research. The Battelle team members regarded their Stony Brook counterparts as unprofessional, mocking the way they evaluated resumes as if they were hiring professors. The Stony Brook team members, on the other hand, feared for their independence and for the academic atmosphere of the lab, and were horrified that the Battelle crew evaluated resumes as if they were hiring businessmen. "It was ugly," said a member of the Battelle team; "We had weeks of strife," said a Stony Brook participant.

The clash was apparent even in the styles of the team leaders. Battelle's CEO Douglas Olesen and PNNL director Bill Madia were corporate businessmen, while Kenny was an English professor. The two Battelle leaders felt that their company's vast corporate structure and lengthy experience with the DOE entitled and even necessitated it to lead any partnership; Kenny, who saw that Battelle wanted Stony Brook to participate in name only, refused to accept any arrangement in which Stony Brook was even a hair less than an equal partner, and eventually forced Battelle to back down and treat Stony Brook more as an equal. Stony Brook and Battelle formed an independent nonprofit entity, Brookhaven Science Associates (BSA), to bid for the contract. It would have a sixteen-member governing board with five members from each of the partners, and six of the nine universities that were in AUI. This was clever, for it kept enough of the AUI universities involved so that none would be tempted to join some other team, while diluting their influence enough to prevent the DOE from concluding that it was the same cast of characters. In early August, BSA had a kickoff meeting for its "Brookhaven Proposal Team" in Stony Brook's Presidential Conference Room before its Stony Brook team members flew to Battelle's Columbus headquarters to complete the proposal.

The RFP required each bidder to specify a director on the bid—in fact, all administrators down to the levels above department chairs—and to commit themselves in writing beforehand. The Battelle team saw Madia, the PNNL director, as an obvious choice. Stony Brook would have none of it, leading to more tension between Madia and Kenny. While Stony Brook prevailed, it would not be easy to find the right person: an eminent research scientist who could manage large organizations but who also didn't already have a permanent job that they were willing to walk away from at a moment's notice to commit themselves to a position that didn't exist yet. Stony Brook's search committee turned up only one person who agreed to be considered for the position as presented: Venkatesh Narayanamurti, universally known as "Venky," an eminent scientist and manager of a strong research group at the University of California, Santa Barbara.

Venky, age 58, had gone to the University of Delhi, received his PhD from Cornell, and worked at Bell Labs and Sandia before joining UC Santa Barbara. He had frequently received offers to head laboratories, but was uninterested in administration and had turned them all down. He was ambivalent about the Brookhaven offer. The lab had a superb reputation in high-energy physics and materials science, but Venky did not want to abandon his research. He had been spoiled by having worked at Bell Labs, which had left him free to pursue his own direction, and had liked Sandia until, after the Tiger Teams, he found it growing bureaucratic in a way that interfered with his work. Venky believed in the value of great research organizations, but thought that they needed a degree of independence and insulation from Washington. He liked to say that "You can't run a lab like a religion, where people have to swear to do things in a certain way."[12]

Venky visited Brookhaven in August, and said he would need an offer guaranteeing him the independence and clout to run a forefront lab, and to ensure that, he asked to be appointed a distinguished professor with tenure at Stony Brook. On August 22, about a week before the proposal deadline, BSA sent Venky an offer letter. Increasingly confident, BSA also announced that it would host a reception at Stony Brook the day after Labor Day, intending to reassure Brookhaven employees about their intentions, the leadership they were bringing, and their employment status. The Stony Brook officials who set up the meeting expected that the lab scientists, whom they all knew, would welcome them with open arms.

But Venky soon felt religion emanating from Battelle and too much bureaucracy from Stony Brook. Battelle seemed interested in doing whatever the DOE wanted, and while Stony Brook said that they would request a distinguished professor appointment, which would be rubber-stamped, the appointment couldn't be made on the spot and had to go through the state university's office in Albany. When Venky received BSA's appointment letter he hesitated, fearing that Stony Brook was unable to push Battelle, the DOE, and Albany hard enough to give him enough independence. On top of that, with the deadline for proposals looming, he was under pressure to sign on the dotted line. As he put it years later, "I didn't want to become the burnt frog," that is, get into a situation where he would be slowly roasted to death.

* * *

At the beginning of August, after a month of amicable discussions—in contrast to what was happening to the Stony Brook–Battelle team—RPI and Westinghouse announced their partnership, and Lahey moved to Pittsburgh to begin hammering out a formal proposal to submit to the DOE. For the rest of the summer he and other RPI team members shuttled back and forth between Troy and Pittsburgh. "The only one who was really happy about this was US Air," Lahey recalled. RPI would be the lead institution, Westinghouse the partner. As the team would be pitching the proposal with an engineering focus, Lahey was the natural choice as the proposed director.

Westinghouse had an experienced proposal-writing unit with its own building across the street from its R&D center. Like many corporate proposal-writing units, the Westinghouse unit was broken up into "Blue," "Red," and "Gold" teams. The Blue team wrote the proposal, and then presented it to the Red team, whose job it was to tear it apart; afterward, the Gold team—largely made of former DOE people who knew how to construct a proposal that would sing for the DOE—put on finishing touches. "It's a night and day operation in which you don't sleep much," Lahey recalled. "We spent more or less seven days a week working until one in the morning." It was brutal. "They were bounty hunters," Lahey recalled of his presentation to the Red team on August 20; "they demolished me."[13] Then the Gold team got to work.

Two hitches remained near the end of August. One was that the Westinghouse proposal-writing unit was simultaneously working on a proposal to bid again for Oak Ridge, causing delays in the final version. The second was that RPI President Pipes was still dragging his heels signing off, and the RPI–Westinghouse team asked the DOE for an extension of the deadline. The DOE was not concerned; bidders often need more time and in this case it was not unexpected given the condensed timeline. The agency extended the deadline from August 28 to noon on September 8, giving them the Labor Day weekend plus another week.[14]

This delay also turned out to be a great relief to the Stony Brook–Battelle team because their proposed lab director had not committed himself, and would not by August 28, the original proposal deadline, meaning that the Stony Brook–Battelle partnership could not satisfy the RFP requirement for a signed-on director.

SCRAMBLING TO MEET THE NEW SEPTEMBER 8 DEADLINE

Both teams utilized the new deadline to good advantage. As the RPI–Westinghouse proposal was wrapping up, Wilson, RPI's Director of News and Communication Lisa Rudgers, and a few others flew to Pittsburgh. The RPI and Westinghouse groups had worked so well together that after they finished, Wilson took them all to dinner at the Station Square Grand Concourse Restaurant, the most elegant in Pittsburgh, which had marble columns and a stained-glass, vaulted, cathedral-like ceiling. The dinner was magnificent, but Wilson made the mistake of allowing oenophile Lahey to order the wine, and the group had the most expensive of the restaurant's already-pricey selection. "I ended up paying out of my own pocket," Wilson recalled, "since I could not turn those bills in to RPI with a clear conscience."[15]

After dinner, Wilson took the team up the century-old cable car to the top of Mt. Washington and its spectacular views of downtown Pittsburgh. "It was a beautiful evening and Larry [Larry Snavely, vice president of government relations] pulled out some cigars and we all smoked a cigar on the overlook as a celebration," Wilson wrote. "That included me, a complete non-smoker who had not even smoked a cigarette before."

Right after Labor Day, the RPI–Westinghouse team sent their Gold team's proposal to the printer, and Lahey, Wilson, and the other RPI team members left Pittsburgh for Troy to make a final presentation, on September 6, to Pipes for his blessing. Lahey also invited Bond to drop by to bring him up to speed. Bond flew up from Brookhaven, and the two met Saturday morning. Before returning to BNL, Bond dropped in on an RPI physicist whom he knew. In the course of the conversation Bond asked if his colleagues were excited about the bid. Bond was shocked to hear the physicist say that he was troubled that though the bid was due in less than 48 hours it had not yet received final approval from Pipes. Bond returned to Brookhaven, baffled.

That afternoon, Lahey and the team ran through the final proposal with Pipes, assuming that approval was perfunctory. Pipes then stunned everyone by declaring that he would not sign—and that he was leaving to Lahey the humiliating task of informing Westinghouse that his university was withdrawing from a carefully worked-out agreement with a global and well-connected corporation at the last minute.

The pull-out was so bizarre that it inspired speculations. Rumors circulated that D'Amato influenced RPI to drop out, wanting Stony Brook

University, with whom he had ties, to win without competition. Others thought that Pipes was concerned about the financial implications—if Brookhaven, say, were to lose class-action suits from neighbors seeking compensation for environmental damage. Still another rumor was that Pipes was concerned about congressional support for the lab given D'Amato and Forbes's efforts to shut the reactor. As *Nature* pointed out, "Most government laboratories in the United States rely on the unstinting support of their local political representatives, without which they are unlikely to win battles with other laboratories for programmatic funding."[16] Not Brookhaven.

Still another factor was baser: a growing animosity between Pipes and the rest of the RPI administration. After Pipes's refusal to approve the bid, the enmity between he and Lahey came to a head, in the wake of which Pipes set out to fire Lahey as dean. There followed a soap opera–like episode that is legendary in RPI history. Wilson brokered an arrangement whereby Lahey would not be reappointed when his term expired the next spring; Pipes initially agreed, then fired Lahey anyway. This was the last straw for the faculty, who in April 1998 passed a no-confidence vote against Pipes. He resigned the next day, thought better of it, and recruited his local congressional representative to help him get reinstated, in vain. Pipes refused to leave the president's mansion and had to be escorted out a few months later.

Around 11:00 p.m. on September 7, the day before the deadline, Westinghouse notified Silbergleid of the bizarre twist that RPI was withdrawing—but that if Silbergleid would give them a two-week extension the company guaranteed that they would have another partner and get in a competitive proposal. The competition now had only one bidder, which was the last thing Peña wanted. Still more embarrassing was Secretary Peña's promise that if the competition did not go as envisioned he would reconsider the entire process. The next morning Silbergleid cashed in on his demand that Peña get back to him within a day. He got through to Peña's office about 9:30 a.m., outlined the situation, relayed the Westinghouse promise, asked for a two-week extension, and held his breath. Half an hour before the deadline Peña's office called Silbergleid giving him the go-ahead to extend the deadline. In public, the agency tried to give credibility to the bizarre step by saying that

Westinghouse didn't need much time to work things out with a new partner given that it had already completed a proposal.

* * *

The Stony Brook–Battelle team faced a different crisis. On August 29, Venky told the BSA team that he was pulling out as proposed lab director. It wasn't as precipitous a decision as it may appear. Given the rush of the competition, he had been first contacted only about two weeks previously, and received an offer only the week beforehand.

The proposal deadline was a week away, and if Stony Brook couldn't come up with a director, Battelle would surely appoint Madia. The Search Committee had wanted to recruit a director in an open search, but now they had no choice but to find someone in their ranks. There seemed only one possibility: Marburger, Kenny's predecessor. AUI had sounded him out as a possible director in March and as a possible interim director in April—and BSA's search committee had contacted him in July—but he had declined each time. Now he seemed BSA's only credible possibility.

Marburger was a solid choice. A physicist and seasoned administrator, he had defused numerous controversies. Born in 1941, he received his BA from Princeton and PhD from Stanford. He joined the Physics Department at the University of Southern California in 1966, becoming department chair. In 1976 he became dean of the USC College of Letters, Arts and Sciences. In that role he was the administration's spokesman in several crises, the first being a scandal involving preferential treatment for athletes. By then he had gained experience with speaking to large audiences in authoritative and engaging ways as host of an early morning TV science show. He knew how to cultivate the impression he made on others. He sometimes advised faculty on how to select clothing to convey particular impressions to specific audiences, often instructed cameramen on how to film him, and gave administrators books about leadership. Marburger had the uncanny ability to calm any meeting and make everyone in it feel heard and valued, even when nothing actually happened. Those who regularly attended his meetings and found they unexpectedly felt good about the meeting's progress learned to scribble down the thing that he had just said. One of his favorites was "You know,

I'm more optimistic than I expected to be." It seemed to work like magic; everyone read a different thing into it and assumed that's what he had meant.

Marburger knew the importance of image control in teaching, governing, and interacting with the government, media, and public in highly charged circumstances. Another of his tricks was to show people the MG he kept running flawlessly and the harpsichord he had built from scratch, and then subtly let them know how to interpret these possessions of his—that he was skilled at making things with hundreds and even thousands of parts work together effectively.

In 1980 Marburger had become president of Stony Brook University. In just three years he had demonstrated such an ability to handle controversies that New York State Governor Mario Cuomo appointed him chairman of the Shoreham Committee "fact-finding panel" to study the safety and policy issues of the by-then controversial Shoreham Nuclear Power Plant. The panel contained a full spectrum of Long Islanders, from antinuclear activists to scientists. Yet Marburger skillfully managed to get all the members of the group to sign their names to the final Shoreham document. Afterward, Cuomo wrote him that "Not since Captain Bligh has anyone commanded a crew trying to sail in so many directions."[17]

Early Monday morning on Labor Day, September 1, Marburger fielded a call from a member of the Stony Brook team outlining their dire predicament: its choice for director had just withdrawn, the proposal was due in a week and a day, and without a director the proposal was dead in the water. Marburger then spent much of Labor Day on the phone with BSA members, mostly to find out if they would understand if he served only a year or so as a placeholder. Three days later he and BSA member Robert McGrath, Stony Brook's Deputy Provost, were on the plane to Battelle's proposal center at its headquarters in Columbus where the proposal was being readied for submission the next day, September 5, so it would be sure to be there by the deadline on September 8.

"It was a gray morning," McGrath recalled later. "After our struggles with Battelle the previous two months I was wondering how Marburger would play out." Not to worry. Marburger and McGrath entered the Battelle building and were led through the bowels of the building and into a cramped room packed with Battelle employees. McGrath introduced Marburger and retired to the back, taking an empty seat beside a consultant. "Jack started talking," McGrath said, "and proceeded to give such an astonishingly clear and compelling talk that it bowled everyone over. The consultant leaned over and whispered to me, 'Who the hell *is* this guy?'"

That weekend, Madia, exhausted by weeks of negotiating with Stony Brook but fully confident of the proposal, left for vacation in Seattle. Meanwhile, on Tuesday after Labor Day, September 2, the Stony Brook–Battelle team held the reception it had promised the previous week to introduce the team to Brookhaven employees. But now, with the bidding deadline unexpectedly extended, Kenny had to answer questions about tenure from anxious Brookhaven scientists tersely with "It's proprietary," leaving few feeling either enlightened or welcomed. Another dampening factor was that the organizers of the reception seemed unaware of the D'Amato and Forbes press conference earlier that day at which the congressmen had called for the HFBR's closure.[18]

On the morning of September 7 Madia was in Seattle at the Four Seasons hotel with his wife when he received a call from Silbergleid—an old friend, with whom he kept in touch throughout the competition—telling him that RPI had pulled out of its partnership with Westinghouse and that the DOE was again extending the proposal deadline to give Westinghouse time to work out a proposal with a new partner. Madia was outraged; it was unheard-of to pull out of a partnership with a company the size of Westinghouse when seeking a $400 million contract with the DOE. The DOE had reset the deadline another two weeks, to September 22 at 1:00 p.m. Chicago time.[19] Madia told the press, "It's like coming up to Super Bowl Sunday, you're ready to play, and somebody comes up and says, 'Let's play next week!'"[20] He did not speak with his old friend Silbergleid for weeks.

From the outside, the DOE's process seemed slapdash, inspiring a spoof that circulated at BNL imitating a *Newsday* article. Entitled "711 and Mattel to Run BNL," with the byline "A. Nonymous," it read:

> Spookesville, September 5, 1997—711, a well-known convenience store chain, and Mattel Inc, the toy company giant, expressed interest in bidding

as the contractor to run Brookhaven National Laboratory in Upton, NY. . . .
[A spokesperson said that 711 has] the expertise to run large facilities and the
close proximity to stores near BNL would make this operation just an extension
of their activities . . . [and] mentioned the possibility of using the cryogenics
facilities to develop new varieties of Slurpee which could be marketed worldwide including the Gobi desert and Atacama desert in northern Chile. . . . Mattel spokesperson, Mr. Ken G. I. Joe [mentioned] that Mattel has the ultimate
experience in developing "things." . . . The consortium will nominate a director
as soon as possible but they stated: "There is no rush to nominate a director. It
is preferable to select a good candidate. In the meantime one of the nearby 711
store managers could take a second job to run the lab as an interim director."[21]

SCRAMBLING TO MEET THE FINAL DEADLINE, SEPTEMBER 22

The day after RPI had pulled out of the proposal, members of the Westinghouse team were frantic. But it had a carefully thought-out proposal,
and the team was confident it could find a replacement for RPI among the
places where its jet had landed six weeks before. One obvious prospect was
the Illinois Institute of Technology Research Institute (IITRI), which had
been interested in bidding all along but had never found an opening. It
had numerous contacts with both Westinghouse and with RPI, thanks to
a facility that it ran in Rome, New York, two hours away from Troy. When
Westinghouse called, IITRI was only too happy to join.

IITRI was founded in 1936 as the research laboratory of the Illinois
Institute of Technology (IIT), though it eventually became independent.[22]
IITRI was nonprofit, had 1,200 employees, operated a number of laborato-
ries in 14 states, and held several contracts with the DOE and other federal
agencies. Its resources included the Illinois Institute of Technology High
Energy Research Group, headed by Leon Lederman, whose Nobel Prize was
earned for work at Brookhaven, while its Center for Synchrotron Radiation
Research and Instrumentation did research similar to that of Brookhaven's
NSLS. But IITRI's prime resource was its engaging and energetic president,
Bahman Atefi. Born in Tehran, Atefi had studied at Cornell and received

his PhD from MIT. He was known for being able to digest huge amounts of material in a short time. He worked under Robert Bari at Brookhaven as a nuclear engineer, then—like almost everyone else in the field—realized that designing nuclear reactor cores was a dead end, and switched to studying the risks of nuclear power plants. After a few years he moved to IITRI and soon became its president. A strong promotor of Iranian-Americans, Atefi had founded the Public Affairs Alliance of Iranian Americans.

While the Stony Brook–Battelle team had had almost two months to put its bid together, the IITRI–Westinghouse team had two weeks. Much of the first week was devoted simply to getting people from the two organizations together to meet and greet. Since IITRI had just joined, much of the bid would have to rely on what Westinghouse had already written. While Westinghouse would provide the industrial backup and coordination of environmental safety, IITRI would be responsible for the scientific research. It would also have to find a director.

On September 8, back at Brookhaven and still shaking his head at the RPI pullout, Bond got a call from an IITRI representative requesting a meeting off-site on Friday morning at which Bond was asked if he would consider becoming the lab director in their team's proposal (as a sign of how fast-paced the events were, that afternoon was the day the Bond–Forbes debate was taped). Bond noted that a bid whose proposed director was the deposed contractor's director did not seem a good strategy. But the IITRI team needed a name immediately, as the bid was now due in ten days. Bond agreed to be a placeholder—satisfying the letter of the RFP though not the intent—and was promptly written in to the proposal being worked on in Westinghouse's proposal-writing building.

On September 20 Bond and a few others from BNL traveled to Pittsburgh. It was two days before submission, and the BNL team had to meet the IITRI–Westinghouse team and discuss how presentations would be made to the DOE. Bond, the presumptive director candidate, had not yet seen the proposal and had no idea of its contents. There wasn't time to make changes and reprint the RFP response by Monday, but they improvised. The IITRI–Westinghouse proposal was delivered on Monday just before the 1:00 p.m. Chicago-time deadline. As per the RFP, it included sixteen copies so that each of the Source Evaluation Board (SEB) members and advisors had their own.

Now that BSA had Marburger on board, its team used the extra ten days to fine-tune its proposal and prepare for the upcoming presentations. It helped that Marburger, unlike the highly preoccupied Bond, was essentially retired, though he continued to hold a faculty position in the Physics Department, and could devote his attention full-time to preparation and visit Columbus several times. Battelle had a "boot camp" at its Battelle headquarters in Columbus to train team members. As Battelle had often bid for DOE contracts, it had the system down, and the camp was run smoothly and efficiently. Blue and Red practice events were set up, with the Blue team giving the presentation and the Red team picking it apart. Other professionals were present to critique the presentations, which were structured as to what had worked for the DOE as possible. Marburger had time to attend all sessions. The new BSA proposal was again dropped off early, on September 19, in a truck driven by a highly ranking Battelle official who was a former DOE manager.

Each team had created a new nonprofit entity to be the prime contractor, as per the RFP requirement. The prime contractor for the IITRI–Westinghouse team would be the IITRI Brookhaven Company, while the prime contractor for the Stony Brook–Battelle team would be Brookhaven Science Associates.

The path to that moment—the culminating phase of a $400 million-per-year, five-year contract—had been chaotic and at times even haphazard. Neither team had a complete proposal at the original deadline in August. By the first extended deadline, one had chosen a director less than a week before the first extension, while the other had changed partners less than twenty-four hours beforehand and now had had only two weeks to make the team cohesive. One of the few people to be a key member of both teams was Marge Lynch, the assistant laboratory director for Community, Education, Government, and Public Affairs. But the DOE was pleased with that, viewing her as an agent of change; she had been less than three weeks into the job when AUI was fired, and had no deep ties either to AUI or to the lab.

The two teams made two rounds of oral presentations to the DOE's ten-member Source Evaluation Board (SEB), one at the beginning of October and the other two weeks later. The DOE presentation process had a regimented, almost medieval and inquisitional flavor. Presenters would go into a room in front of a line of committee members who had been trained to sit and listen impassively, without laughing, frowning, or showing any reaction. The presentations could not be written, only oral, and presenters had to speak directly to the sentence or clause. Both Battelle and Westinghouse knew the system, and tuned their presentations accordingly.

The BSA team gathered at a nearby hotel on September 26, the weekend before its orals. During breaks, the conscientious Marburger retired to his hotel room to practice his oral presentation, while the self-assured Madia returned to his to watch the Ryder Cup, the biennial golf tournament taking place in Spain that weekend. BSA had the system down, and the weekend went smoothly.[23]

Bond flew to Washington to prepare for the IITRI–Westinghouse presentation on September 27. The team was in some disarray: many of the team members had not met each other, and there was no time to make changes. Presenters were handed viewgraphs of material they had never seen before. On Sunday night Bond had to fly back to Brookhaven to deal with issues including how to address the discovery of radioactive water in the BGRR air ducts. He flew back on Tuesday afternoon in time for the presentation the next day, October 1, held in a conference room in a Washington hotel. Already busy as the lab's interim director, he had little time to work on team building or the presentation.

A few days later Silbergleid faxed the SEB feedback to each bidder. The panel was unhappy with one of BSA's leadership choices, but after discussions its choice was approved. The six pages of questions and concerns faxed on October 7 to the IITRI–Westinghouse team included a more serious issue: they expressed concerns about Bond. The key part of the statement questioned if Bond "had the leadership qualities necessary for championing the desired ES&H and community culture change, for bridging the cultural differences between existing staff and the new Westinghouse management philosophies."[24] There was also a hidden subtext: with Bond as the team's proposed director, the DOE would be in a bind as to whether the team won or lost the contract. If it won, the lab would have the same director as under AUI, which was sure to make D'Amato go ballistic again and give the activists more fodder. If the team lost, Bond would still be an interim director approved by the DOE for an additional four months and would lose credibility as having been rejected by the DOE. The panel's statement convinced the IITRI–Westinghouse bidding team that a change was needed despite the short timeline. In a few days they had managed to find and convince someone known to IITRI and DOE, and one who had also worked at BNL: Michael Knotek.

Knotek was a solid choice, an eminent scientist and seasoned administrator who had worked at Brookhaven and with the DOE. From 1985 to 1989 he had been chairman of the NSLS, and had managed to improve machine performance and turn it into one of the country's most productive user facilities. In the process he came to know Brookhaven, its strengths and weaknesses. He also knew Battelle as in 1989 he moved to the Pacific Northwest Laboratory (PNL) and helped the transformation of that laboratory into a national laboratory, renamed the Pacific Northwest National Laboratory (PNNL), with a new user facility, the Environmental Molecular Science Laboratory. After seven years he left PNNL for the Advanced Photon Source at Argonne National Laboratory, just outside Chicago.

Bond was at home on the morning of October 11 when he received a call with the news that he was being replaced as proposed director, though he was also told that they wanted him to remain on the team as the deputy for science and technology. Meanwhile, he was due to leave the next day for Pittsburgh to prepare for the second round of presentations on October 12–13. Events had moved so swiftly that the news about Knotek had not gotten around to most members of the team, who were shocked. With only a little over a week to go before the final SEB presentation, Knotek had to be incorporated into the bid quickly. "It was a short fuse," Knotek recalled years later.

At the second round of presentations four days later, at the DOE's Chicago Office at Argonne, Knotek gave a knowledgeable presentation of the science and capabilities of the lab, and of the community atmosphere and activists around BNL. At lunch, a panel member stopped by the table where Bond was sitting with others and brought up the subject of UFOs, and said that he had concluded that UFOs were a result of quantum fluctuations of the universe. Bond wondered, "And this person is playing a role in determining the management of BNL?"

Following both panel presentations the SEB panel went into deliberations for a month—twice as long as the IITRI–Westinghouse team had spent writing its proposal. To most outside observers, BSA seemed the clear favorite, given that Stony Brook was nearby and the biggest user of BNL facilities, while Battelle was prestigious and well known. But the IITRI–Westinghouse team had pulled itself together in the intervening two weeks, and Knotek had given a superb presentation. "Lots of people, particularly in the Office of Science, assumed that it would be a slam dunk for BSA," Silbergleid said. "But the competition turned out to be a lot closer than anybody thought it would be."

The first week in November, the BSA team had a shock: Kenny announced that she was a short-listed candidate for president of the University of Texas at Austin, her alma mater, and would probably accept if selected. The BSA's proposal had been finished and submitted; still, she had been a key figure in its shape and success, and had almost single-handedly prevented Battelle from dominating by establishing Stony Brook as an equal partner. She was also slated to be BSA's first chair, to rotate between the Stony Brook president and the Battelle CEO. Some Stony Brook officials were worried about the partnership breaking apart if Kenny left.

The process was also jolted by the discovery that the recently enacted Energy and Water Development Appropriations Act contained a procurement rule that required the DOE secretary to issue a waiver prior to awarding a "noncompetitive" management and operating contract—here, that the bidders were restricted to nonprofit entities—and also required that the secretary had to notify Congress of any such waiver at least sixty days beforehand. A few days into November, Secretary Peña decided to interpret this rule as applying to the Brookhaven case.[25]

This delayed the transition another two months. There would be a sixty-day period prior to the new contract signing, and another sixty-day transition period afterward. The earliest date for a contract signing was now January 5, 1998, and the earliest date for the takeover was March 1, 1998. Associated Universities, Inc., would be in charge an additional two months, or four months after the DOE's original date of November.

Brookhaven employees learned the news of the additional delay on November 5, the day Peña notified Congress. Many were suspicious. Some assumed the secretary's unexpected action indicated he had not been sufficiently interested to pay full attention. Others assumed that his narrow interpretation of the procurement rule reflected pressure on the DOE by for-profits—and by Brown and Sensenbrenner—angered at the restriction to nonprofits. Still others thought, conspiratorially, that the DOE was planning to scuttle the competition and use the narrow interpretation as a chance to reopen it. Marburger wrote, "There is a little skullduggery and a lot of chaos out there . . . a lot of finger pointing and complaining. . . . The DOE has given the enemies of not for profit operation of BNL an opportunity."[26]

Annoyed at this unexpected twist, D'Amato and Forbes pledged to find a way around the requirement. Though they did not manage to advance the contract signing, they did succeed in moving up the ability to announce the winner.

DECISION

A week later, on November 13, the panel reported its evaluation, which favored BSA. Very early in the evaluation process one of the panel's members had stated that only one team had knowledge of BNL's science and was a clear preference. The Department of Energy had settled on BSA and gave them a "two-minute warning," allowing BSA to prepare a press release, "Battelle, SUNY–Stony Brook Team Wins Brookhaven Contract,"[27] as well as a flyer introducing itself, complete with a BSA logo.[28] But the formal announcement was put on hold for a final DOE review.

The evaluation document praised both bidders, but a key factor for the choice of BSA was "its clear superiority compared to IITRI–Westinghouse in the area of Scientific and Technological Programs," which was the highest weighted of all evaluation criteria. This was largely based on their having more experience in the areas of BNL's science mission. In the areas of operations IITRI–Westinghouse was viewed as being stronger in a number of areas, including the all-important ES&H; Westinghouse's corporate experience indicates "a broad scope of recognized ES&H leadership and record of accomplishment at DOE facilities." Both teams showed they could effect the "necessary cultural change," but BSA had more experience in relevant research areas. IITRI had no experience in nuclear physics, and only limited experience in high-energy physics. BSA therefore was rated "outstanding in scientific programs management," with the IITRI–Westinghouse team rated "satisfactory." Both were said to have "outstanding" proposed lab directors. "For the reasons set forth above, it is my determination that the selection of Brookhaven Science Associates provides the best overall value to the government."[29]

The official announcement came on November 24.[30] That evening, after a heads-up, the BSA team gathered in Kenny's conference room at Stony Brook, put on the speakerphone, and received the formal news that it won the contract.

Most major players in the competition wound up happy. Forbes and D'Amato obtained their environmental credentials, Battelle established a solid claim to nonprofit status, Stony Brook eventually joined the AAU, and Westinghouse ended up not having to recompete its West Valley contract in 1999.

In a prepared statement Peña called it "the fastest competition we have ever had for a management and operating contract and reflects a new way of doing business at the Department of Energy," and said that it was "open, fair, and efficient."[31] That's not how it had seemed on the inside. In light of the chaos that the expedited bidding generated, the "new way" shortly reverted to the "old" length of time used to compete contracts.

6 END OF A TURBULENT YEAR

We'll close the lab down . . . that's what God would intend.

—HELEN CALDICOTT

CONTRACTOR-SELECT

November 25, two days before Thanksgiving, came the public announcement that BSA had won the contract to manage Brookhaven, effective in another three months. BSA members met with BNL administrators and addressed an overflow meeting with lab employees. The new upper management included two people from Stony Brook, including Director John H. Marburger III and five from Battelle.

Krebs opened the meeting by introducing Marburger, who began by saying that he would like to personally thank Peter Bond for his role in holding the lab together in the past stressful year. Marburger's words evoked a standing ovation in the overflow crowd—a spontaneous and unexpectedly deep emotional moment for Bond and for the audience. Only when the extended applause died down did Marburger introduce the BSA team. BSA head Kenny amused everyone again with her line about how she was the first Stony Brook president who was not a white male physicist named John. She then made a few comments about perception, reality, and Brookhaven's problems over the past year.

Nearly everyone at the lab was happy with the result and especially that the arduous process was finally over. Stony Brook University officials called the decision a reflection on its preeminence as an educational institution. "It

Figure 6.1
DOE Office of Energy Research head Martha Krebs welcoming John H. Marburger III, whom she had just introduced as Brookhaven National Laboratory's new Director-Select, at BNL's Berkner Auditorium, November 25, 1997.

is a rare institution in this country that has this kind of responsibility and the quality to carry it off," said Kenny proudly, pointing out that only two other schools in the country run national labs, the University of California at Berkeley and the University of Chicago (both in the AAU). "That's nice company."[1]

Local politicians were pleased. D'Amato, who had lobbied hard to have AUI fired, called the outcome "magnificent."[2] Local urban planners saw benefits. "It is a triple A positive for the economy," said Lee Koppelman, the executive director of the Long Island Regional Planning Board. "I can't think of better news in this Christmas season. It's a great gift for Long Island."[3]

Robert L. Park mockingly put things into his perspective by noting that "the minuscule tritium leak from the reactor at Brookhaven, which led to a change in management, is nothing compared to leaks at the Hanford reservation on the other side of the continent, where 67 of 177 storage tanks are leaking really bad stuff. Traces are now reported to be showing up in groundwater. The concern is that it could eventually reach the Columbia River. Battelle, which now worries about leaks at Brookhaven, is a Hanford contractor."[4]

A few days after the announcement, *Newsday* reporter Charlie Zehren met Director-Select Marburger for breakfast at the Three Village Inn in Stony Brook. Marburger arrived in his 1970 MGB GT, a striking two-seated classic sports car that he maintained himself. He liked to off-handedly drop that fact to reporters for them to use as a symbol of his skill in the art of making highly complex machines (metaphorically meaning those involving people, not just parts) run flawlessly. (Marburger's handling of the Shoreham Committee was a much more salient example of his relevant skill, but the sports car image made for a better news angle and photo op.) Zehren took the bait, or most of it. *Newsday* illustrated his article with a picture of Marburger half-standing and half-sitting on the car, leaning into it. He's

Figure 6.2
Brookhaven Laboratory Director-Select John H. Marburger III, posing with his MGB GT for *Newsday*. From Newsday. © 1997 Newsday. All rights reserved. Used under license.

wearing a coat and tie, with one hand on the car and one hand on his thigh, and is looking straight at you with a stony, confident face and half-closed, determined eyes—you trust this man. The article's lead sentence referred to the MGB GT as "the perfect car for a man who's trained to plumb the mysteries of the universe." But while Zehren dutifully picked up Marburger's cue to use his maintenance of the car as a symbol of his leadership skills, he tried to use the car as a symbol of the lab itself, referring to it as unresponsive and rumbling.[5]

Marburger promptly wrote *Newsday* a correction. My car "runs like a watch," he wrote. "It doesn't smoke, has no electrical problems, and has plenty of power for its size and age." It may seem churlish to be concerned about such a tiny correction in a positive article, he continued, but these kinds of assumptions get us into trouble. Marburger said that he did not want employees to make assumptions about potential safety hazards, nor did he want the lab's neighbors to make assumptions about the safety of air or water. He will give complete and accurate information. "This time the story was cuter and no one suffered from the inaccuracies," Marburger concluded. "Next time please be more careful."[6]

Newsday did not print the letter. But that Marburger wrote it at all is a sign of the way he seized opportunities to cultivate an image to help him cope with the tasks he faced—in this case, to reinforce the perception of his own competence and trustworthiness.

A few weeks later, another *Newsday* reporter interviewed Marburger at a local Italian restaurant. At one point the journalist asked: "Do you believe in God?" It was a dangerous question. The movie *Contact* had just been released, in which the character played by Jodie Foster was interviewed by a commission to judge if she was fit to be selected for an interstellar journey. Thrown by the identical question that Marburger was asked, Foster beat around the bush before eventually replying—reluctantly and defensively—in the negative. The answer caused the commission to decide against her, although she later gained the position after a convenient plot twist.

Marburger, however, was being posed the question on Long Island, not Hollywood. He knew that there would be no plot twist to the rescue should his answer prove controversial. He loaded spaghetti onto his fork, popped it into his mouth, and decided to spend a long time chewing while he thought about how to answer. Given the chain of events that had led him to become director, Marburger realized the danger of appearing at odds with the values of the external community. As a former university president, however, he was also aware of the danger of diminishing his authority inside the institution should he pander to those values. He knew that it would have been acceptable to say something like, "That's private matter and I don't feel comfortable talking about it in public." But mindful of the urgency to convey openness and honesty, he decided to engage the question straightforwardly but carefully.

"I don't adhere to any particular organized religion," Marburger said. "I believe there are mysteries in the universe that we don't understand yet, and perhaps never will. I grew up in the context of the Methodist Church and it helped me a lot. I have nothing but affection for the church. The question of believing in God and supernatural forces is difficult for me to answer because I understand so much about physical forces in the universe and so little about human forces."[7] That response, which *Newsday* duly printed, showed that Marburger was willing not only to field questions from reporters about everything, including his thoughts on the divine, but also to take the time to give a thoughtful reply. Marburger's honest answer—stating uncertainty but also a cautious inclination toward the negative—expressed respect for, and even made use of, the power and legitimacy of values that he acknowledged were not his own. His response was a brilliant example of speaking with integrity while reaching out to the values of those who considered him an opponent.

Back at the lab, several scientists made light of the fact that their boss-to-be was the sort who not only was willing to field questions from reporters about the divine, but took the time to give a thoughtful answer. It was certainly not something that any previous lab director would have been likely to do. In some respects the incident was trivial, but it reflected a major change in the way a lab director responded to the public.

Another illustration of the new style Marburger brought to the position took place after he received a letter about the HFBR from a local middle-school student named "Franklin," who was concerned about the HFBR. My friends, Franklin wrote, "would like to know your beliefs, the ideas you have about this issue and how to solve it." Marburger responded seriously and thoughtfully. No radioactive contamination from the reactor poses any threat to health or the environment, he wrote. "People who have been reading the headlines would probably not believe such a statement, and yet that is the conclusion of all the government agencies that are charged with protecting public health and environmental quality." But, Marburger continued, let me go through the issues systematically, so "you will get the flavor of how scientists view these things." He wrote out for Franklin what was known about groundwater and aquifers, nuclear reactors, nuclear reactions, nuclear fission, nuclear fuel, the HFBR's emission of contaminants, the reactor's normal operation, the spent fuel pool and its leak, the safety of radiation, environmental and health impacts, and the political and social factors in the way of a restart—hundreds of words about each topic, written in language a middle-schooler could understand. At the end Marburger wrote:

> Whom can we believe about these things? I will close by telling you how I decide. I tend to pay more attention to people who try to make distinctions between large and small effects than to people who make sweeping statements on either side. I listen to people who seem to be willing to change their minds based on new information. I am skeptical about the views of people who use dramatic or accusatory rhetoric to make their points. I tend not to listen to people who stereotype "scientists" or "bureaucrats" on the one hand, or "environmentalists" or "activists" on the other. I value humility in the face of ignorance, and courage when the facts are clear. Above all, I never give up trying to understand everything from my own knowledge and learning enough to make sound judgments without relying on the opinions of others. Science is made by ordinary men and women, and scientific knowledge is not reserved for some special class of people. I believe that if something is understandable at all, then all should be able to understand it.[8]

The letter, and its conclusion, is an outstanding exhibit of Marburger's communicative skills. He says that he is effectively in the same place as Franklin—confronted by a mixture of chaotic voices and the problem of who to listen to—and shares his counsel about how to decide. That final paragraph outlines how the HFBR's fate should have been decided, and how it wasn't.

Peña had fired AUI based on what he said was AUI's loss of community trust, and hoped that the incoming manager—with Marburger as the incoming lab director—would quiet the activists, restore community trust, and normalize relations between the lab and politicians, neighbors, and media. No such luck—or not soon. For the militant activists it didn't matter who was in charge or even how competently the lab was run. "It's a sham," said Bill Smith. "With John Marburger heading the new group, the attempt will be to perpetuate the nuclear agenda."[9] Roger Snyder, another activist, said, "The new faces had ties to BNL. How can they be any different?" For the activists, all that mattered was to close the HFBR.

"Under the new management," *Newsday* editorialized on the day before the Thanksgiving break, referring to the manager-to-be, "let a fair assessment begin."[10] Not a chance.

ALEC BALDWIN

STAR, the new antinuclear group crystallized by Baldwin, quickly overshadowed and marginalized other civic and community groups. It was interested not in starting dialogue with the lab but in stopping it. STAR saw only two positions that anyone could take toward the HFBR and even the lab—unqualified support or closure—and viewed the latter as right and the former wrong and corrupt. STAR would make life as difficult as possible for groups that thought otherwise.

Baldwin had a house in Amagansett, one town farther east on Long Island than East Hampton and about forty-five miles east of Brookhaven. At something of a lull in his film career, he was flirting with political ambitions. On November 24, the headline on the cover of *New York Magazine* was "See Alec Run: Tired of Hollywood, Alec Baldwin Is Preparing for the Role of His Career: Candidate." He told the reporter that "I want to be a 'feral' Democrat" and a "ferocious liberal," and said he could "beat the crap" out of Forbes if he wanted to, though he had still higher ambitions.[11] "Alec Baldwin does not want to be Senator, Governor, or President . . . Yet," ran another article, and Baldwin informed the reporter that the Brookhaven plume was so dangerous that it was "on its way here, to the East End of Long Island."[12] Baldwin was unhappy when a writer for *New Republic* followed him around for a day and wrote a less than fully flattering profile. Baldwin had fulminated about Brookhaven and its reactor that, he alleged, was near his house and leaking radioactive waste, and drew a sketch of the plume for the reporter, over which he drew an arrow supposed to point to East Hampton (though the arrow was not pointing in the direction the plume was traveling): "Every year, the radiation is coming closer to my home." The reporter fact-checked, and learned that Brookhaven's reactor was used for research, that the plume had not left the lab site, and that in any case the plume was not moving anywhere near the direction of Baldwin's home in distant Amagansett. The reporter also discovered that Baldwin had declined invitations to visit Brookhaven and inspect the reactor himself.[13] Baldwin soon changed his mind about running for office. "Well, it was a fun story while it lasted," editorialized *Newsday*.[14] Instead of becoming a feral Democrat, Baldwin became a feral anti-Brookhaven activist.

The antinuclear activists now had a hotheaded celebrity, a well-funded organization, a six-figure budget, first-rate legal advice, a professional PR firm, Democratic Party influencers, and access to national television shows. It held fundraisers in art galleries in the Hamptons and luxury apartments in Manhattan, raising $156,550 in 1997 and aiming for $1 million in coming years. It had an office in East Hampton with full-time staff upstairs from an art gallery, just above where the luxury real estate brokerage office of Sotheby's is now.

But the group's most important asset was Baldwin. Celebrities are a boon to causes. They are charismatic, able to motivate and focus people, and know how to enthrall reporters and beckon cameras. Baldwin, in particular, knew how to get audiences to ignore personal warts and transform a bad boy reputation into an engaging, even cute persona with a social conscience. His crude and belligerent language and outrageous statements were charming; when challenged, he might walk back his remark, or say he was misquoted or joking. He rambled in public pronouncements, but his celebrity status and powerful media presence attracted new audiences.

The formation of STAR was a new stage in the effort to close the HFBR. The activists could now promote their cause not just through pieces in *Suffolk Life*, *Rage*, and *Newsday*, but through national talk shows, major magazines, and society columns. They recruited attention and contributions from other celebrities, including Barbra Streisand (known for her support for progressive causes), who got their knowledge about Brookhaven from Baldwin.[15] ("Say it ain't so, Barbra," the BNL progressive Mark Sakitt said when he read of Streisand's contribution.) Instead of hanging signs at highway overpasses, STAR members could hire airplanes to fly banners over Long Island beaches reading "BNL + H_2O = CANCER." Late in the summer at an event in the Hamptons, Baldwin ran into TV host Montel Williams, who volunteered to help the STAR cause.

No incident at the lab was too trivial to be overlooked as a sign of potential disaster. When, one hot summer day in March a few years later, some hay caught fire in an open field, not involving any radiation, STAR's lawyer Cullen viewed the lack of an independent investigation as a red flag. "If it happened once, what's going to prevent it from happening again?"[16]

"[T]he formation of East Hampton-based STAR . . . represents big new trouble for B.N.L," wrote Grossman. "It signifies the shift to the national arena of the controversy over Brookhaven National Laboratory and the radioactive pollution it generates."[17]

STAR claimed for itself the role of representing the "community,"[18] and scorned other groups with broader goals and less abrasive conduct. The LIPC was one; Judy Pannullo continued to plan programs and dialogue with the lab. In the fall of 1997 the LIPC described its aim as continuing "to function as the eyes and ears of the neighboring community," and listed three ambitions: (1) to ensure that "all potential health effects from BNL of concern to the local community be monitored by researchers independent of BNL and DOE"; (2) to convince BNL to "consider research topics which are of particular interest to the local community, such as breast cancer and alternative energy," as well as global warming; and (3) to expand community oversight.[19] The organization established a committee to investigate the lab that it called "Neighbors Expecting Action and Remediation" (NEAR)—to emphasize its desire to work cooperatively with the lab and the fact that its members were locals rather than wealthy members of the East Hampton enclave. After a STAR member accused LIPC of "going to bed with the enemy," Mannhaupt retorted: "We are the immediate stakeholders at hand. We have the backyards that butt up to the lab . . . ten years from now, when all those little groups are gone, we'll still be here."[20] When both STAR and NEAR competed for an EPA Technical Assistance grant, and NEAR won, Baldwin called up Pannullo to curse at her. To calm herself, Pannullo, who was on crutches following an accident, sat down and wrote a poem in which she contrasted the blast of heat that had singed her, sent by a "star," with the glint of silver light from her crutches.

The LIPC members deeply resented the media coverage STAR was getting and were repelled by its confrontational tone and sole focus on shutting the HFBR. STAR's goal, they felt, was not to determine if the reactor was safe, still less to determine how to make it safe, but just to get rid of it. "All of a sudden, from the left wing, come people like Alec Baldwin," Mannhaupt said. "They are not schooled enough, technically or scientifically. You can't just stamp your feet and get what you want."[21] (Baldwin and STAR would prove her wrong.) LIPC members and other local environmental groups began to refer to the STAR people as "Gucci carpetbaggers."

STAR not only attacked civic and community groups who wanted to dialogue with the lab but also attacked all studies and data that did not declare the lab to be endangering the community. While contamination of the Peconic River had been studied earlier by New York State, STAR and Bill Smith pushed for additional testing. "Why would you believe the data they were generating?" Smith said. "We didn't believe the data because it was their data."[22] In an effort at community outreach and collaboration, the DOE and Suffolk County signed an agreement with Smith's Fish Unlimited to split samples for analysis. Yet when it was time to compile the various analyses for cross-checking and verification for a report to Suffolk County, Smith did not submit his.

STAR held a high-end fundraiser on October 17 in a posh English manor–like luxury mansion at one end of Main Street in East Hampton; at the other end were parties and screenings for the Hamptons International

Film Festival. Attendees were expected to donate between $500 and $5,000 for the crusade "against toxic nuclear radiation from Brookhaven National Laboratory."[23] Featured speakers were Baldwin, Caldicott, and Jan Schlictmann. The activists hired a prominent PR firm, Fenton Communications, to promote the event.[24] At it, Caldicott called for the entire lab to be closed.

This was too extreme for Forbes and D'Amato, who said they were unequivocally opposed to closing the lab; still, STAR's prominence helped to exert political pressure.[25] Its efforts brought more politicians to the bandwagon or solidified their stances about the lab. One was New York State Attorney General Dennis Vacco, up for reelection, who on the day of STAR's opening benefit issued a document calling for "Major Brookhaven Lab Reforms." A contingent of BNL and DOE officials (Helms, Davis, Bond, DOE lawyer Joan Shands) met with Vacco's staff to correct some of the charges of the report and update changes that had taken place, and Helms wrote up a response that the DOE headquarters lost for six months, to Vacco's aggravation.[26]

After reading Baldwin's "Environmental Suicide" letter in the *East Hampton Star*, in which Baldwin had suggested that many of Long Island's health issues are due to Brookhaven, Bond invited Baldwin to visit the lab. Baldwin "formally accepted," canceled, accepted, and canceled again.[27]

INDEPENDENT REVIEWS

The year had brought many reviews by outside agencies. The November GAO report commissioned by Sensenbrenner was released.[28] The forty-six-page document, "Tritium Leak and Contractor Dismissal at the Brookhaven National Laboratory," reviewed the full story—the agreement to install the monitoring wells near the reactor, Paquette's 1994 recommendation to install more wells, the lab's failure to do so because of low priority and lack of funding, the eventual installation of the wells in 1996, and the first results indicating the presence of tritium in the groundwater in January 1997. The GAO report mentioned the pre-1997 tests that showed the spent fuel pool was not leaking, and the 1997 ones that did. The report discussed the DOE's inspections and reviews of the lab, and the events leading up to the DOE's decision to fire AUI.

"GAO report on tritium leak spares no one," commented Park in *What's New*. It criticized the lab for delaying installation of the wells and not living up to its commitments to Suffolk County. It criticized the DOE's local office for not aggressively monitoring the lab's progress in meeting the commitments. It criticized the DOE program offices—the Office of Nuclear Energy, Science, and Technology; the Office of Energy Research; and the Office of Environmental Safety and Health (headed, respectively, by Lash, Krebs, and O'Toole)—whose poor integration and confusing and overlapping responsibilities "prevented effective accountability." The report included Schwartz's admission that AUI had overlooked "the potential political impact" of a leak at the spent fuel pool, but also Schwartz's charge that "the Secretary was poorly advised by his senior managers" as well as Schwartz's bitter observation that "Our MSIP was incorporated in its entirety by DOE into its plan for improvement of management of BNL and will now become the benchmark for whichever new contractor is selected to manage BNL," demonstrating that the DOE could have achieved its desired outcome by encouraging rather than firing AUI, "with far less cost to the Government and trauma to scientists at BNL, users of its facilities, and the scientific community in general."

Still more damning, the GAO report pointed out that the DOE itself had known of its own integration and accountability problems since 1993, when the DOE had reviewed its oversight of the lab and the GAO reviewed the DOE's management of the national lab system. "DOE's management structure problems are long-standing," the GAO report concluded. "[D]espite many calls for improvements by internal and external groups, DOE leadership has so far been unable to develop an effective structure that can hold its laboratory contractors accountable for meeting all important departmental goals and objectives."

The GAO report criticized Peña's decision to fire AUI. All senior DOE officials interviewed by the GAO investigators agreed that the tritium leak itself was not a serious health hazard and that the problem concerned "the public's perception of the way AUI managed the problem" and "the loss of the Long Island community's trust in BNL." Yet the GAO report found that, until 1996, the DOE did not evaluate AUI at all on public trust issues—and that the DOE's own performance measures did not reflect the priority it now said it was giving to environmental safety and health issues. Further, in the DOE's 1994 and 1995 appraisals of the lab, the DOE did not specifically identify ES&H criteria, which were "relatively minor aspects of each year's evaluation." In fact, the GAO noted that the DOE had to look beyond "mere mechanical application" of its performance criteria when it retroactively downgraded its rating of BNL's performance of 1996. The GAO report suggested that this may have been a reason why the DOE terminated the contract for "convenience" rather than "cause"—to avoid "a possible legal challenge by AUI over performance criteria."

The GAO report prompted Sensenbrenner and Brown, the chair and ranking member of the House Science Committee, to write to the DOE asking why not much had changed since the earlier reviews critical of the agency. "This report," Sensenbrenner wrote, "indicates that the failure to discover the tritium leak in a timely manner was due, in part, to systematic and management problems at DOE," he said. "This raises serious concerns about DOE's management and environmental compliance at other DOE labs."

The GAO report also unlocked some censored internal DOE documents. In September, AUI had filed a breach-of-contract claim against the DOE for terminating its contract. The early termination, the claim said, would cost the AUI $8.1 million in lost fees, and the claim noted that the DOE had "intentionally misled" AUI by grading it highly on its evaluations. The claim also said that the contract called for problems to be handled "in the spirit of friendly cooperation" and that the DOE's action was "a clear abuse of discretion and compelling evidence of bad faith." For example, Secretary Peña had refused to meet with AUI officials prior to his decision to terminate them, and the DOE's own ISME had called the lab's actions "aggressive and appropriate." The claim went on to say that the DOE's actions had undermined AUI's ability to attract potential bidding partners and thus its ability to function as a viable organization. Besides monetary damages, the claim also mentioned contractual issues involving benefits due employees and the AUI's relationships with its participating universities.

Along with the claim, AUI filed a Freedom of Information Act (FOIA) request for documents associated with the termination. The DOE delayed compliance, claiming executive privilege for some documents. In November it released some heavily redacted documents and withheld some entirely—nearly all of the Options Paper was censored. The GAO investigators, however, had had access to all the documents, and the DOE was eventually forced to hand over the unredacted documents discussed above. Both the deleted passages and the justifications that the DOE gave for withholding them were embarrassing. The deletions include passages about the desirability of using AUI's firing as a tool for sending "a strong message" to other labs with "the beneficial effect of demonstrating DOE's seriousness," the fact that the DOE was setting about downgrading AUI's management rating, and the looming presence of D'Amato. The deletions also showed that the DOE considered postponing firing AUI for six months in order to avoid having to pay BNL employees severance pay. The deletions had been made less to preserve confidentiality than to protect the DOE's reputation.

Other reviews by such independent organizations as the DOE, ATSDR, EPA, and eventually Suffolk County and New York State focused on environmental issues and concluded that environmental issues were under control and were not health hazards.

The DOE Office of Oversight of the status of Groundwater Tritium Plumes Recovery concluded: "Current management of the BNL tritium remediation project is effective and progress has been substantial. Continued attention is needed to ensure that ongoing activities are completed on schedule in the face of a number of upcoming challenges such as: potential funding and staff reductions, the upcoming transition of contractors and the need to devote resources to issues identified by the environmental vulnerability assessment and other groundwater contamination discoveries."

The ATSDR study found low levels of an array of chemicals in wells south of the lab boundary, but "none of the chemicals was in great enough concentrations to endanger people who drink the water." The report continued, "The levels of individual volatile organic compounds (VOCs) and radionuclides are not sufficient to produce adverse health effects." Environmental professionals respected the ATSDR report. James Pim of the Suffolk Health Services Department said, "The report does nothing but reiterate the known facts . . . and their conclusions are reasonable."[29]

The militant activists promptly denounced the ATSDR, a federal agency. "Whatever they're saying . . . there's a deeper issue here," Maniscalco said, without specifying what it was.[30]

The independent Suffolk County Environmental Task Force, led by the Stony Brook epidemiologist Roger Grimson, concluded in its preliminary report that no link could be found between the lab and cancer and that the incidence of rhabdomyosarcoma in Suffolk County was low. The Suffolk County Water Authority (SCWA) declared, based on analyses by independent certified labs in Wyoming and North Carolina, that "no sample from any Water Authority well has ever contained tritium or any other radionuclide from BNL."[31]

The tritium plume remediation continued to pump drinking water (with generally less than 2,000 picocuries per liter) from the aquifer, at a rate of about 450 liters per minute, and once the spent fuel vessel was emptied at the end of the year the plume would diminish on its own—but the lab was paying a price for this unnecessary activity, as the total cost of the tritium remediation activities were estimated in April to be $20.7 million, of which BNL was supposed to pay $10.2 million, though over the course of a few months the total estimated cost to the DOE dropped a little and the BNL contribution rose to $11.2 million.[32]

It appeared that there was not likely to be any more explosive revelations. The tritium plume was fully characterized, not moving off-site, and never a threat to begin with, and the one new discovery of radiation-containing water in the BGRR ducts was being cleaned up. Small traces of radionuclides were measured in the Peconic River, but according to New York State officials these also posed no health threats. Apart from occasional discoveries of small amounts of contaminants in the continuing legacy study, not much more would be forthcoming. Brookhaven would not be a Love Canal, Woburn, Toms River, or Hinkley. There would be continuing issues, but no Big Environmental Story—only little ones. Nevertheless, the antinuclear STAR activists now began to focus on preventing the HFBR from restarting by seeking to stop the restart decision process that Peña had initiated.

HFBR AND THE EIS

In the first week of September, containers were loaded for the fourth and final shipment of spent fuel; by the end of the month, all spent fuel rods had been shipped off-site, and preparations were made to add a stainless steel liner to the pool. A press release for the event said that Peña would make the decision about an HFBR restart in four months, by January 1998 (evidently either unaware that he said he would await the EIS first, or of how long the process would take). The next task would be to remove the equipment in the spent fuel pool and empty it of its 68,000 gallons of water.

On November 24, the DOE published a Notice of Intent, which described the HFBR EIS as considering four alternatives: (1) no action, or simply keeping the HFBR shut down; (2) resuming operation at 30 or 60 megawatts; (3) resuming and enhancing its operation; and (4) permanent shutdown and decommissioning. The notice specified the public scoping period as from its publication to January 23. Three public meetings would be held during that period, the first on December 10 and the other two on January 10 and 15. The draft EIS was scheduled for summer 1998, and the final one for November 1998—resulting in a postponement of Peña's decision by almost a year.

This announcement of the restart process intensified the activities of both proponents and opponents of the restart. Proponents set out to make sure that the EIS process laid out by Peña continued as planned; opponents sought to block it before it began. Forbes failed to understand the process, thinking that the EIS was the beginning of the restart itself, and went into a tirade, accusing the DOE officials of not providing "honest input," of staging "dog and pony shows" in preparation for restarting the reactor, and of illegally using taxpayer money to do so. Meanwhile D'Amato refused to let Senator Pete Domenici (New Mexico), a powerful supporter of science who sat on the Committee on Energy and Natural Resources and the Appropriations Subcommittee on Energy and Water Development, to visit the lab to see for himself. In early November Domenici's chief staffer called Bond to arrange an under-the-radar visit on December 3: no publicity. When D'Amato got wind of the visit he insisted on accompanying Domenici, and in the end both canceled. The staffer called Bond to apologize.

STAR collected signatures on petitions to close the HFBR; Friends of Brookhaven to keep it open. Some community groups, such as the Manorville Taxpayers Association, representing a town that borders on the lab, officially supported restart;[33] others from farther away insisted on closing the reactor. The rhetoric became more vehement still, with one antinuclear activist writing that "the overall mission of BNL is to promote the art of nuclear warfare, hence death-oriented. (This was the original mission of BNL and has not changed.)"[34] Both are untrue.

On December 10, the day of the first EIS public meeting, Peña announced that he was delaying the decision to restart the HFBR for a year, to December 1998. *Science* quoted an unnamed lab official saying what everyone thought, that this step was motivated by the impending reelection campaigns of D'Amato and Forbes. "Peña does not want to give them this issue to run on." The decision date would also be after Peña left office, which he had been thinking about but had not yet made public.

FOB PIVOTS

On hearing of the STAR fundraiser, Friends of Brookhaven invited Baldwin to meet with them. "We know of no credible scientific studies that link Brookhaven to cancer clusters," studies made by federal, state, or local regulatory agencies, the invitation letter said, pointing out ATSDR's recent conclusion that the lab's contamination is "not sufficient to produce adverse health effects."[35] The FOB members said they wanted to discuss the evidence with the actor and, independently of Bond, invited him to the lab as well.

Meanwhile, FOB members tried recruiting their own celebrity. They approached Alan Alda, host of the popular science television show *Scientific American Frontiers*, but Alda's staff informed them that the issue was too controversial for their boss to get involved. FOB also tried to forge political relationships, and through union connections contacted John Powell, the head of the Suffolk Republican Committee. Shady but effective, "Mugsy," as he was nicknamed, was a legend on Long Island, having skyrocketed from highway department payloader driver to becoming, at age 35, one of the most powerful politicians in Suffolk County. Powell told FOB that he could help, and wanted to work out an arrangement. One Friday afternoon after hours FOB members met with Powell at the lab.

It did not go well. "We need your help," Kaplan said. Powell was abrasive. "Every time we Republicans look at Brookhaven," he said, "we see 5,000 Democrats," saying that he wanted to change that, and would help if the group set up a table in Berkner at lunch hour so that lab employees could register Republican. As Powell went on and on with his message, some FOB members drifted out; one began mocking Powell behind his back. Powell left annoyed.[36] The next year he was arrested and then convicted on federal corruption charges.

Then came a turning point. The lab's media relations department sponsored a course in crisis communication by Vincent Covello, a professor of environmental sciences and clinical medicine at Columbia University and founder of the Center for Risk Communication. Though initially afraid that the course would be "psychobabble," John Shanklin and others signed up. "It changed my life," Shanklin said.[37]

Covello began the course by reminding the scientists of a lesson in persuasion two and a half millennia old. Persuasive speaking, the Greek philosopher Aristotle had outlined, appeals to the audience in three ways. One is through the speaker's credibility and character, another through the soundness of the argument, and the third by connecting with the situation of the audience. In Brookhaven's case, Covello told them, this translates to the following: establish your credibility by citing scientific sources, clearly express what the evidence points to in terms that nonscientists can understand, and engage your audience. Scientists, he said, do the first well, can sometimes do the second, and are extremely poor at the third.

The first thing you *don't* say, Covello taught, is something like, "The tritium's not a problem, we've analyzed the numbers, don't worry about it." The first thing you *do* say is something like, "This should never have happened, we are fixing it, and as members of the community we too are concerned." You also *don't* call what the activists are doing a "witch hunt," for it creates a barrier between "us" with good values and "them" with bad ones.[38]

FOB members now presented themselves very differently at public meetings—as neighbors rather than experts. Post-Covello course, as they walked in the door of a public meeting they would say things like, "Hi, Jeremy, glad you came out!" or "How are your kids?" so the audience would trust them at least a little by the time they said things like, "I work at the lab and the story about it in the papers is nothing I can relate to based on what I know." This sharply contrasted with activists' tone and their accusations that the scientists were conspiring against the community. Jennifer O'Connor was particularly effective, for she was able to say things like, "I work at the HFBR; do you think as a mother of two young children that I would do anything to endanger them?" Shanklin recalled, "We knew we wouldn't get the activists on our side, but as we left meetings people began to say, 'I had a feeling it wasn't like *Newsday* presented it but I didn't know all the good things that the lab does. Why doesn't the press report on that?'" The group also began to attend Forbes's town hall meetings. Reading back his public statements, they would ask respectfully that he either back them up with evidence or come to the lab to get the facts for himself.

"We were relentlessly positive," Shanklin recalled. "We wanted to act as the reasonable, accommodating, nonthreatening presence." Unlike many colleagues, the FOB members were determined to be collegial and friendly even with the activists who were treating lab employees as pariahs and denouncing them to the press as murderers. When Maniscalco held one hunger strike at the lab entrance, one angry employee driving back from a local McDonald's threw a hamburger at him. An FOB member, on the other hand, greeted Maniscalco at a public meeting and made him break character and laugh by saying, "Pete, you're looking great! Have you lost weight?" After members heard Bill Smith say that he had had a heart attack, they tried to find out if he had had a thallium stress test, which would allow them to ask what he made of the fact that part of what had saved his life had come directly out of the lab.

Advertising at STAR's lavish level was out of the question. Many lab employees supported National Public Radio, which gave those who contributed certain amounts of money to fund drives the opportunity to have something read on air; FOB asked lab employees to give to FOB instead, to be bundled into a single lab contribution to WSHU, an NPR member station. They managed to raise enough to have "BNL: We care about our neighbors, our environment, and our science" read on air several times per day.[39]

After STAR collected 2,400 signatures on a petition to close the HFBR, FOB circulated its own petition calling BNL "a valued asset to the Long Island community" and that the future of the HFBR and its research should be decided "only after completion of the ongoing evaluation."[40] Unlike the activists, lab employees generally had no experience with large-scale signature-gathering, and FOB had to send out reminder emails begging people to "swallow their fear" and seek out "strangers and business owners," for "THE THREAT TO OUR LABORATORY IS REAL!! The anti-Lab/ HFBR activists have already gathered thousands of signatures against us. Thousands." The email continued, "The anti-Lab/HFBR activists are counting on our apathy. I for one am not going to roll over for them."[41] FOB ultimately gathered over 18,000 signatures. STAR charged that the signers were all lab employees and their families and friends. Rowe countered: "My husband and I collected 40 signatures. Two of these were my husband and myself, but the rest were gotten at places like a soccer game, talking to other parents at my daughter's karate practice, and with some members of the orchestra I play with. These were not family friends."[42]

STAR organized a "call-in day" to the office of Patrick Moynihan, New York's other senator, to involve him in their cause; the FOB countered with a call-in campaign of their own. "There is only a handful of activists but there are over 3,000 of us," their flyer declared.[43]

FOB members also revamped their communications with journalists. Journalists, Covello had pointed out, know how to elicit heated and extreme remarks from people, remarks one can come to regret. Every time an incident took place that might have journalists calling, FOB members would get together and think about the message they wanted to see in the newspapers, and then would type it out on a piece of paper and tape it to the phone. When a journalist called, they then read what was on the paper—and if the journalist asked other questions, hoping to provoke an impolitic burst of anger, they would simply read it again.

While Bond kept his distance from FOB, having to interact with the DOE, Forbes, the press, and the community, he kept periodically informed of their activities. "Double-Oh-Seven here," Bond would answer when recognizing Shanklin on his caller ID. "It was a light touch that helped keep our spirits up, no matter how bad things were," Shanklin said.[44]

The FOB members went to extraordinary lengths not to be confrontational. One day an antinuclear group notified *Newsday* that they were coming to the lab to chain themselves to the fence. Shanklin showed up and explained that, as an open research center, Brookhaven had no fence—but added light-heartedly that the Friends of Brookhaven knew people at the lab who could quickly make a section of fence for them so they could go ahead with the interviews as planned. "It was straight out of Monty Python," Shanklin said.

"THE LAB REPRESENTS EVIL"

But FOB members were unprepared for how extreme STAR members were becoming.

Ten days after the announcement that BSA had been awarded the contract to manage the lab, STAR sponsored a public meeting called "Brookhaven National Laboratory: Problems and Possibilities." Speakers would include Caldicott, an economist and business professor at Colorado College named William Weida, and Jay Gould, a statistician and economist, whom STAR members had adopted as an authority on radiation effects, though he was regarded with skepticism and even amusement by the mainstream scientific community.

In *The Enemy Within: The High Cost of Living Near Nuclear Reactors*, Gould—an antinuclear activist with no training in epidemiology—asserted that low-level radiation caused AIDS, Lyme disease, breast cancer, infant mortality, low birthweights, low SAT scores, and other ailments. The book was ridiculed by epidemiologists and health physicists for faulty approaches, unsupported claims, and portentous statements such as that "some malevolent force of mortality is being emitted from reactors."[45] Gould was also convinced that Sr-90 was being emitted by both Millstone and the HFBR, and started a project to collect baby teeth to analyze and prove it, placing advertisements in local papers asking for baby teeth. Health scientists at

Brookhaven and elsewhere in the country scoffed, saying that such a study is inherently flawed for a number of reasons, including that the environment already contains trace levels of Sr-90, that the preponderance of dairy products that contain it are not from Long Island but from places like New Jersey, Upstate New York, Wisconsin, and Vermont—and finally that Gould was anointing himself to oversee the study.

Gould often partnered with Ernest Sternglass, a retired physicist who claimed that radiation caused high infant mortality and crime rates, low SAT scores, and an abundance of other social and health ills. BNL scientists "don't give a damn about human life," Sternglass said.[46] Gould and Sternglass once claimed that higher-than-normal breast cancer rates in towns within fifteen miles of Brookhaven might be linked to radiation releases from the lab and that towns downwind of the lab had cancer clusters. They had mislocated the lab, assuming that it was in the South Shore hamlet of Brookhaven rather than Upton, a low-cancer-incidence region according to their own data. When the lab's location was plotted correctly, the *lowest* cancer incidences proved downwind and downstream of the lab.[47]

The STAR-sponsored public event place took place on a snowy evening, December 5, at Guild Hall, an arts, entertainment, and education center in East Hampton. It was supposedly educational, but no lab employees were allowed to speak except from the floor. Schlichtmann had been advertised as the moderator, but he did not appear, and his replacement, Alice Slater, began by reporting that the lab has "no inventory of toxics so we don't know what is there," adding that "we must get rid of the nuclear weapons at BNL"; other speakers also mentioned that BNL had nuclear weapons. Helen Caldicott mentioned doomsday scenarios, including disasters caused by chemicals from potato farming, radiation from the Millstone nuclear plant in Connecticut and its impact on the food chain, release of the BGRR's radioactive cooling water (the BGRR was air-cooled), and plutonium supposedly dumped on the BNL site. The "dangerous synergism" between all of these things, she continued, meant that "$1 + 1 = 10$."

Weida spoke about the need to convert BNL from a weapons lab, and claimed that it was spending very little money on cleanup. He distributed a document that outlined many of STAR's claims about BNL, including that the lab was principally engaged in nuclear research, implying weapons research. Bond, who attended, found the document. "I picked up a copy of your December 4, 1997 study on Brookhaven National Laboratory at the STAR meeting," Bond emailed Weida a few days later. "It has minor as well as major errors of fact—where did you get your information?"[48] Weida's answer: "crawl back into your hole." Bond responded by pointing out more errors, starting with the fact that Weida had gotten the (easily discoverable) acreage of the lab wrong by a factor of almost half and that the cleanup funding was not in the documents Weida had looked at. "Nice try, but no cigar," Weida replied. "You know, you guys really are not very good at this. I hope you are better at doing your other jobs."

Weida passed the exchange on to STAR board member Roger Stanch-field, who cited Bond's corrections as yet another example of Brookhaven's "passive-aggressive intimidation" by hinting that "big brother is watching." Stanchfield wrote, "My hat is off to you for telling Mr. Bond where to go."[49]

A few weeks later Bond, as requested, sent STAR his corrections to Weida's report. No, the lab does not and never did have nuclear weapons. No, the amount spent on cleanup is not $5 million but $29.5 million; no, it is untrue that the lab will spend nothing on remediating pollution for the next five years; no, it is not true that the lab spends $300 million on nuclear research—that's the budget of the DOE's Office of Energy Research, which covers biology, high-energy physics, nuclear physics, and environmental science; no, foreign scientists don't siphon money from the lab but bring it in; and other corrections.[50]

Bond's list of corrections did not go over well, and prompted Weida (evidently unaware that Bond would be out of office in about six weeks) to seek to have Bond fired. Bond, Weida wrote DOE Secretary Peña, "is incapable of interacting with the citizens of Long Island" in a cooperative manner, and Peña should "remove individuals of this ilk" from the lab's management.[51]

Two days after the Guild Hall event, protestors blocked the lab entrance, leading to the arrest of eight people. "The lab's conduct is more criminal than what I'm doing," one participant said.[52] Just over a week later, on December 17, STAR organized a rally and brought Catholic Bishop Thomas Gumbleton from Detroit to urge the closure of the HFBR and the laboratory itself.

At the lab gate, Gumbleton led a group of activists in prayer and speeches, and justified his opposition by citing the evil of military research and the right of religion to override science. "I see immorality here," he said.[53] Just as the Catholic Church can be definite about its opposition to abortion because science is indefinite about the moment a cell becomes human, so morality can be definite about opposing the lab because science is indefinite about the health risks. "The human person is so valuable that you cannot do anything that risks killing a human person."[54]

Gumbleton was followed by Helen Caldicott:

> This lab represents evil. . . . They're pumping water incredibly contaminated with strontium-90 and plutonium out of that lab tomorrow. The water is in 60,000 gallons of insulation pipe and I have evidence that it was put there purposely by the lab. . . . Don't lie to us. Tell us the truth. We own this lab. We pay for it. It's owned by the people—not to kill the people, but to help us, and this lab's killing people. . . . So they're pumping the tritium plume at the southern limit of the plume, pumping it and sucking the most concentrated water down toward the populated people—populated areas of Shirley and Brookhaven and the like.

What the lab was really doing was the opposite, pumping the water *away* from the people down south, taking the edge that was going toward the lab border and sending it back. Caldicott then mocked Friends of Brookhaven and statements in some of its literature that radiation can be used for the benefit of humanity, such as in treating cancer patients and that the radionuclide americium is used in household smoke detectors. No, Caldicott told the demonstrators, Brookhaven scientists are not curing cancer but using that as a front for "playing with their nasty nuclear stuff." As for americium, she continued, it's "more dangerous than plutonium, the most deadly substance on Earth." She concluded: "And the lab knows what it's doing. And what it's doing is evil. Thou shalt not kill. . . . We'll close the lab down, and the people can be re-employed to do things to save the planet. That's what God would intend."[55]

Long Island's Catholic leader was less sure of the Church's doctrine on scientific research than Detroit's Gumbleton. Bishop John R. McGann, the leader of the 1.3 million Catholics on Long Island, said he was reserving comment until he saw the Environmental Impact Statement.[56] Catholic employees at the lab, who included Doug Ports, also responded to Gumbleton: "I am appalled that a leader of my church would make such irresponsible, cold, and uninformed statements."[57] Ports then composed a reasoned defense of the lab citing the US Catholic Conference's monograph, "Renewing the Earth," a summary of the Church's scriptural teachings on the stewardship of God's creation: "Scientific research and technological innovation must accompany religious and moral responses to environmental challenges," the monograph said. "Reverence for nature must be combined with scientific learning."[58] But Gumbleton's words and pictures, not those of McGann or Ports, made the newspapers.

"WHORES" AND "HOLLYWOOD REDNECKS"

After having received two invitations from Bond to visit the laboratory, and agreeing and canceling, Alec Baldwin finally agreed to meet with laboratory officials—but off-site and accompanied by other STAR members.[59] Geiger coordinated with STAR on the topics they said they wanted to discuss, including health studies, the BGRR, the HFBR, RHIC, employment, and the status of the plume. The meeting took place on December 22 at Stony Brook. Gunther, Lynch, Helms, Marburger, and Bond met with STAR representatives Baldwin, Cullen, Schlichtmann, Peter Strugatz, and Grossman. Cullen told the press, "The purpose was to have a private discussion and not an opportunity for anybody to grandstand." As a letter to *Newsday* later put it, "Yeah, right. That's why they brought along a movie star."[60]

The meeting became newsworthy even before it began. Standing outside the meeting room beforehand, Alec Baldwin noticed *Newsday* reporter Jordan Rau. Baldwin demanded that Rau not attend what he called a private meeting between STAR and the new management, with no press. Rau asked why he was prohibited and not Grossman, who called himself a journalist but who had spent two decades writing negative articles about Brookhaven. Baldwin said that Grossman was attending as a member of the group and not as a reporter. Carrie Clark said that Grossman was there in an "advisory capacity." Cullen wrote that there was no media present.[61] Grossman said,

"I was there as a journalist and an observer," but did not object to Baldwin's characterization of him as a member of the group. The STAR members had covered their bases: Grossman was there as a reporter, advisor, observer, and group member, as needed.

The Stony Brook and Brookhaven people sat on one side of the long conference table in the president's office, the STAR people on the other. Schlichtmann started the meeting off on a positive note by saying that STAR wanted to help the community to answer questions and that they wanted "a partnership with the lab." But the other STAR members quickly changed the tone. "I remember that meeting well," Helms recalls.

> Baldwin and Caldicott screamed and yelled and called us crooks and liars and cancer-causers, and used words that would singe your hair. It was absolutely terrible. We kept our cool as best as we possibly could. Baldwin sat right across from me and he kept looking me right in the eye and poking his finger right at my nose and saying "We are going to *shut you down!*" with other foul words spiced in to that statement from time to time. "We are going to *shut you down!*" He said that about a dozen times. I let all that go, and finally I said, "Before you shut us down I'd like to make an offer." He kinda looked back, looked at me and said, "What kind of offer?" I said, "I'd like to offer for you, and whoever you'd like to bring along with you, come to the lab. I'd like you to bring some scientists and technical people, who have credentials, who could sit across the table from our scientists and technical people, and discuss whatever issues you want to bring to the table. We will open our files, we will be as open and forthcoming as we can possibly be. But if you'll bring your people and we'll bring our people, we'll sit down and we'll talk for as many hours as you want to talk. But I want you there, I want Helen Caldicott there, and anybody else, and we can have a third party come and facilitate the discussion if you would like that, and if you want to name a facilitator we can do that too." He [Baldwin] sat back, and he had *nothing* to say. I said, "Just name your time and we'll be there." And he wouldn't say anything. So I said, "Are you refusing my offer?" And he said, "I'll get back to you." He never did. Not one word.

Caldicott was even more vehement. She said that STAR knew the lab had hidden the leak for twelve years and that the BGRR had been leaking its tritium-containing cooling water. When Bond corrected her, pointing out that the BGRR was air-cooled, she began screaming at Bond, accused him of lying, and said that she thought the BGRR had been closed after a nuclear accident. Baldwin had to intercede to calm Caldicott down, the first of many times. "She went a little bit nuts at the meeting," Marburger said later.[62] Baldwin said that BNL needed more oversight, whereupon Bond went through the list of independent agencies that already had oversight over its operations: the EPA, the DOE, Suffolk County, New York State. The STAR members then said that they wanted to study the lab employees' health records; since they were convinced that the lab had significantly increased cancer rates on faraway places on Long Island, they felt sure that the cancer rates of BNL employees most of whom lived within a fifteen-mile radius of the lab—must be even higher.

After Rau wrote in *Newsday* about being barred from the meeting, Baldwin labeled him a "whore."[63]

The following morning, Marburger, Bond, Rowe, Lynch, and Helms met with *Newsday* managing editor Howard Schneider, reporter Zehren, and a few others. The purpose was to address *Newsday's* accuracy and coverage of the lab, and the lab officials brought along examples of what they thought were the newspaper's mistakes, mischaracterizations, and inflammatory headlines, as well as its failure to cover significant stories. But the conversation inevitably turned to the meeting the previous day. Baldwin was "incredibly misguided about the facts of BNL & radiation," Marburger said. Caldicott "relies on histrionics even more than Alec does." Grossman "was himself," quoting from one of his articles and giving Marburger a copy. Marburger said that STAR was a "small group" with "disproportionate influence because of $$$ & visibility of Alec Baldwin." STAR wanted to demonstrate "their ability to get BNL to the table. They are not interested in "fairness" or in being corrected."[64]

To an interviewer years later, describing the meeting with Baldwin, Marburger let slip his carefully cultivated neutrality. "I hated the guy; he's a rude, crude loudmouth," Marburger said, "a typical Hollywood redneck."[65]

NEW YEAR'S EVE 1997 AT BNL: LEAK STOPPED, SPENT FUEL POOL EMPTIED, SPIRITS UP

Despite the STAR events, lab officials had reasons for optimism as the year drew to a close. The lab's scientific activities and RHIC construction were proceeding well. The environmental news was good: the sewage treatment plant upgrade was completed on December 17, and the state-of-the-art, $13 million waste management facility opened the day after. Lab scientists had developed advanced environmental protection methods to improve handling of chemical and radiological waste associated with research programs and other operations. The spent fuel pool was being emptied, as was the BGRR air duct, and the water shipped to Savannah River, and plans to reline the pool were proceeding. The second phase of the historical vulnerability review was completed along with plans to remediate. The full review had been a major undertaking—400 existing structures, 300 demolished buildings, and 300 portable structures had been systematically and thoroughly evaluated for potential environmental impacts. The lab was taking strong steps toward forming and operating a community advisory council.[66]

By the end of 1997, too, the lab's environmental efforts were firmly under control and certified as safe in studies by such independent organizations as the DOE, ATSDR, EPA, and eventually Suffolk County.

Federal, state, local institutions and others had concluded that Brookhaven had not dangerously contaminated the neighborhood, and nor were there significantly higher incidences of cancer on Long Island. The episode was, as Gina Kolata put it in the *New York Times*, the "Epidemic That Wasn't."[67]

On January 23, 1998, the Suffolk County Environmental Task Force, headed by Grimson and including some STAR members, released its final report confirming what it had found in its preliminary report. "[T]he Suffolk County Task Force on Brookhaven National Laboratory found no elevated cancer rates in communities near the Lab," said Grimson, a biostatistician and nationally recognized expert on cancer clusters. The report presented its evidence that the incidences of rhabdomyosarcoma were lower in Suffolk County than in the neighboring Nassau County and in Queens, Brooklyn, and the rest of New York State.[68] Said Grimson, "There is no cluster of rhabdomyosarcoma in Suffolk County or in the lab." Nor was there a higher rate of breast cancer.[69] The extensive epidemiology report looked at many cancers and malformations with a focus on areas up to fifteen miles from the lab. There were four main conclusions:

1. "[C]ancer rates of all types of cancers are not elevated near BNL."
2. "[T]here is no evidence that rates among the four [directional] sectors are significantly different from each other or are correlated with underground plume or wind direction."
3. "[T]here is no evidence that childhood rhabdomyosarcoma incidence is elevated in Suffolk County or in the circle encompassing BNL during the study period 1979–1993 for which the Registry rhabdomyosarcoma data were available at the time of the request."
4. "[T]he age adjusted female breast cancer rate on the 'east end' of Long Island is significantly elevated (129 per 100,000 women) compared with rates of other areas," which "is not attributable to BNL."[70]

Its final report, issued a year later, would find that cancer rates were not elevated near the lab nor was there evidence of an elevated rate of rhabdomyosarcoma in Suffolk County.[71]

Undeterred, STAR said it needed to study the rates among BNL employees, for an increased cancer rate must be there. The study started to be done, but the data showed that lab employees had no increased cancer rates and in fact were generally healthier than members of the surrounding community. STAR, again refusing to accept contrary data, said that because lab employees have better health benefits than others on Long Island the comparison was invalid.

Another positive sign, oddly, was the outcome of the year-end safety reviews of the lab by the EPA and DOE. Each assessed AUI fines for previously reported events, which the AUI, with some justification, read as a vindication. The EPA investigators had been at the lab for months, and their criticisms were directed as much at the DOE as AUI. In keeping with the strangeness of the events of the previous year, the letter from the EPA was undated and addressed to Robert Hughes, who had been replaced a year earlier.[72] The DOE's investigators had strong incentives to find a Big Event to help justify the decision to fire AUI, and their report cited AUI for "significant regulatory violations of nuclear safety." Yet the violations were for earlier minor safety issues that had taken place and were reported in June: the incident involving two particles of Co-60 on the clothes of workers who were moving dummy fuel elements, the irradiation of the Saran Wrap at the BMRR, and a failure to qualify technicians properly. These had already been addressed and were unrelated to tritium; moreover, AUI was exempt from fines because of congressional protection.[73] The DOE had to combine three post-facto small events to warrant a press release. The agency had to admit that the "actual consequences of the violations were not serious," as well as the fact that the lab had publicly

Figure 6.3
High Flux Beam Reactor, spent fuel pool, empty.

reported them at the time, following established procedures. Still more embarrassingly, the fines were for events that had taken place after the firing of AUI, during a time when Peña and O'Toole were assuring people that the DOE had taken matters in hand.

Park called the DOE's fine for minor safety violations a "Christmas greeting" from the DOE. "The infractions, which were unrelated to the tritium leak, would have incurred a civil penalty of $142,500, DOE said, except the contractor was exempt under the law. Yawn!"[74]

In December, too, the lab buried a time capsule on the occasion of its half-century, containing cotton and pea hybrid seeds, an employee salary schedule, computer equipment, and two bottles of fine Long Island wine. The lab also issued its usual end-of-the-year press release detailing how much it spent on Long Island during the year; in 1997 it was $33 million.[75]

Most momentously of all, on December 31, 1997, the spent fuel pool was fully emptied, except for a small amount to cover the sediment at the bottom to prevent it from becoming airborne; that water would be removed in the coming weeks. Late that day, Bond sent an email to lab employees with that news.[76] The leak discovered at the beginning of the year that had triggered the storm of controversy now was permanently stopped.

The activists' quest for the permanent shutdown of the reactor was unaffected. Maniscalco held a prayer vigil at the lab entrance in the name of the Great Spirit, during which he passed around a foot-long peace pipe filled with tobacco, bark, and berry leaves while a "sundancer" performed.[77] He also asked the local DOE office for permission to have "a full-blooded Navajo Indian" conduct "a Native American Pipe Ceremony at the High Flux Beam Reactor" on the winter solstice.[78] Helms tried to meet Maniscalco halfway, explaining that for security reasons the ceremony could not be held at the HFBR, but could be held outside the gate. Maniscalco said he was going to go on another fast to stop any restart of the HFBR.

All through 1997 protestors had been at the gate almost daily, but as the year wound down their numbers dwindled. On December 31, as the sun was setting on the lab, on AUI's management, and on 1997, a lone protestor stood duty at the gate: Pete Maniscalco.

7 NO RESPITE

Cancer rates of all types of cancers studied are not elevated near BNL.
—Report to the Suffolk County Legislature from the Brookhaven
National Laboratory Environmental Task Force

CONTRACT SIGNING

At 8:45 a.m. on January 5, 1998, representatives of the Department of Energy and Brookhaven Science Associates (BSA) gathered in a small conference room in Berkner Hall to sign the new contract. As many of those present knew little about the lab, Bond reviewed Brookhaven's mission, structure, facilities, and programs, as well as its environmental issues and how they were being addressed. At that moment the management of Brookhaven National Laboratory began to evolve from AUI, the organization that had operated the lab for half a century, to BSA, an entirely new and hybrid corporate-academic organization created for the sole purpose of managing the lab.

The fact that the DOE had shortened the normal transition time from ninety to sixty days made life as difficult for the trio of institutions involved—BSA, the DOE, and BNL/AUI—as had the speed-up of the competition, with the three teams having to work long hours on a grueling six-day-per-week schedule. Each divided itself into over a dozen different subgroups to meet with counterparts to address finance, human resources, community engagement, retirement programs, environmental safety, and the other numerous issues involved in running a national laboratory.

When Secretary Peña terminated AUI, the *Newsday* headline was "Breach of Trust," echoing Peña's statement that BNL had "lost the public

Figure 7.1
Contract signing, January 5, 1998. Left to right: John H. Marburger III, Dean Helms, John Kennedy (DOE), Greg Fess (BSA), and Joan Shands (DOE).

trust . . . because of the confusion and mismanagement that has been going on for years." In the spring, the newspapers had carried almost daily headlines of the tritium leak, but as the year went on many civic groups had supported the lab and the HFBR restart, and had written to Peña to that effect. By "community," the DOE now had in mind the loudest and most unrelenting voices. Here, too, the DOE failed in its intentions. It cited as an example of its success in making a "substantial improvement in community relations" an agreement to allow Fish Unlimited to partner on sampling for contamination in the Peconic River. But that organization, effectively the one-man operation of Bill Smith, was not only unsympathetic but adamantly opposed to the new manager and to the DOE.

Despite the Suffolk County Task Force and studies by other independent agencies, and the DOE's claim of "substantial improvement in community relations," STAR continued its activities just as vigorously as ever, capitalizing on all the resources Baldwin now provided them. In the meantime, Baldwin had recruited several more Hamptons-based celebrities, including the actor Spalding Gray. A few days after the contract signing, Baldwin and STAR made their most sensational and widely publicized performance piece yet on *The Montel Williams Show*.

MONTEL

Before STAR, activist events such as the July 4 protest picnic at the lab gate and the posting of banners on overpasses had been small-impact stuff. Baldwin's celebrity networking and media access now brought anti-BNL activism national visibility, and made it socially engaging and morally exciting. Baldwin sought out his show-biz friend Montel Williams for help. A motivational speaker and Daytime Emmy Award winner, Williams had a tabloid talk show often criticized for promoting pseudoscience and quack medicines, especially after Williams began sponsoring a psychic named Sylvia Browne who claimed that her "readings" could tell parents about their missing children.[1] Browne, who claimed to have seen heaven and angels, told more than one set of distraught parents that their missing child was dead, only to have the child turn up alive—causing one outraged person to open a website called "Stop Sylvia Browne." Still, the show had a national audience, and Baldwin arranged for STAR to appear on it.

The Montel Williams Show set STAR on a different course than Schlichtmann's idealistic image of a movement that took its bearings from scientific studies. "The strategy was that we were going to take advantage of everything we could until they listened," Cullen recalled. "Not everybody thought it [the show] was a good idea."[2]

The show's producers had contacted the lab saying that they wanted to do an episode about its research and asked to interview several scientists, having them sign an "Appearance Agreement" stating that they would not sue the producer no matter how their remarks were used.[3] Suspicious on learning that the show was to involve Alec Baldwin, Rowe arranged interviews with a handful of women scientists, most of whom had raised children at the lab. When all the planned interviews were done, the show's producer asked to have someone talk about what the lab was doing to clean up the reactors. Tempted to do the interview herself, Rowe instead called in Gunther, head of environmental remediation. He, too, had brought his children to the lab frequently. The *Montel Williams* crew also filmed him, then left.

The show aired on January 9, 1998. On it, Baldwin falsely claimed that "the rates of cancer are two hundred to three hundred times the national average in this area on Long Island," while more disinformation came from Williams, who said that "the incidence of breast cancer in Nassau and Suffolk is the highest in the U.S." That wasn't all. Caldicott falsely accused BNL of designing nuclear weapons and declared that New York State was hiding the data that would prove increased cancer incidences near the lab.

The show's theatrical centerpiece was "Kenny," an eight-year-old child with rhabdomyosarcoma. Even if the disease had been linked to radiation, it was physically impossible for plumes from the lab to have flowed to the South Shore where Kenny lived. Nevertheless, Baldwin and STAR members proclaimed that Kenny's cancer, and those of others in Suffolk County mentioned on the show, were caused by Brookhaven National Laboratory.

Baldwin and Williams then used the sick child in a heart-wrenchingly manipulative way.

"I know a lot of adults have said to you what they think has caused this. Why do you think you have cancer?" Williams asks the youth.

A bit trembling, the child obediently answers "Brookhaven Lab."

Williams then tells the child that he will let him star in a public service video. Baldwin says excitedly, "You know the way this works. We'll give you your own trailer!" Williams cheerfully adds, "And your agent can negotiate the fees, all right?" The eight-year-old sounded unenthusiastic, perhaps baffled by being treated as a movie star and addressed in show-biz lingo.

Baldwin suggested that the lab refused to let him enter, not mentioning the numerous invitations he had accepted and canceled. He said that "nothing would make me happier" if independent research showed that the lab posed no cancer threat, not mentioning studies that had shown just that by the US Environmental Protection Agency, the federal Agency for Toxic Substances & Disease Registry, the New York State Department of

Environmental Conservation and Department of Health, and the Suffolk County Water Authority.

The Montel Williams Show was terrific theater staged by veteran show-business personalities who knew how to use the media to deliver misinformation with an emotional punch to a national audience. The Suffolk County Task Force and the New York State Cancer Registry had shown that the rhabdomyosarcoma incidence was not higher around Brookhaven,[4] while according to the American Cancer Society "there are no proven lifestyle-related or environmental causes of RMS, so at this time there is no way to protect against these cancers," though some of the gene changes can be inherited from a parent.[5] Yet Baldwin, Caldicott, and Williams, without mentioning sources, confidently proclaimed to the audience that the rate was much higher around the lab, and that the specific cause was Brookhaven National Laboratory.

The producers of the show did not include any of their interviews with lab women and chose to use only Gunther—and as a foil. After showing a brief clip of him talking about taking his children to the lab, Caldicott said, "They're gonna die." Gunther later had to explain this remark to his unnerved son.

No one who has seen that show forgets "Kenny." No one fails to be moved by a child with cancer. Baldwin and STAR had staged a riveting performance. As a lab employee wrote, "Who could not love a guy who seemingly rides to the defense of unfortunate children with horrendous diseases and does the White Knight routine?"[6] He and other FOB members who watched commented that the STAR performers were manipulating rather than helping such children by assigning blame for their disease to preselected villains while ignoring the best existing knowledge about cancer incidences and cures.

The Montel Williams Show got wide attention, and was seen by an estimated nine million people across the nation; it provoked hundreds of calls to STAR. Liz Smith gave a "Bravo" to Alec Baldwin in her *Newsday* society column. STAR fielded hundreds of calls from people eager to sign on. The lab received horrified letters and emails from people who had seen the program, and had assumed that Baldwin's accusations were true. One two-sentence email asked, "What is being done about the sick children?" to which a lab employee responded with a two-page letter explaining the rhabdomyosarcoma and other cancer statistics on Long Island.[7] The show was designed to spread the message that rhabdomyosarcoma was caused by radiation—and by Brookhaven. At the end of the show, Williams made a slip of the tongue, referring to STAR as an acronym for "Standing for Truth *Against* Radiation."

Friends of Brookhaven members and other laboratory employees were appalled, some comparing it to the accusations at the Salem witch trials and to the rants of Senator Joseph McCarthy.[8] "We need to respond," emailed an FOB member.[9] Another member arranged for the show's producer to accept comments in response, writing, "This is not just about our jobs. I doubt any one of us would put our jobs before the health and welfare of our children. This is about fair play and callous disregard for truth and decency at the expense of an institution with a legacy of benefit for mankind."[10]

The group scheduled a letter-writing campaign, then thought better. "[O]ur letters will be used to fuel controversy and provide material for a 2nd show . . . we have no reason to believe that future shows would be better in terms of fairness to BNL."[11] A few local newspapers published outraged letters. "To allow guests and Montel Williams to mislead the public in such serious, misguided, misinformed facts is criminal," went one letter to the *Three Village Times*. "Don't confuse me with the facts," the *Village Times* editorialized. "But who wants to be bothered with inconvenient reality when pushing a few panic buttons will generate far more publicity?"[12] Another was irate that STAR members had made false accusations of danger, said blatant untruths about innocent people, and made claims of events that aren't happening and of conspiracies that didn't exist: "Let's stop hunting for witches."[13] *Newsday* noted that Williams and Baldwin were promoting "unproven links" between rhabdomyosarcoma and lab activities and that "repeated studies by federal, state, and local government groups and independent agencies have not found any link between the lab's activities and cancer."[14]

Stephen Dewey wrote King Kullen's executive vice president Brian Cullen about his nephew's "insulting, unsubstantiated, ill-founded, undocumented, and reckless statements about the laboratory without even taking the time to learn the facts." Kullen's VP was unimpressed. After Dewey announced that his family would no longer shop at the store, Cullen said it was "unreasonable" to punish the store's employees "because of the opinions of one young man." Perhaps you're right, Dewey shot back, but "Scott's actions, however, are unconscionable, irresponsible, and baseless in their facts."[15] Dewey wrote Rowe, "I am doing a slow crash and burn."[16]

Some Long Islanders detected in the event a downward and dangerous movement in the social standing of science. If influencers on popular media programs could publicly denounce reputable institutions like the EPA and ATSDR, they and their followers could doubt anything, believe anything, and threaten a democracy that depended on the guidance provided by facts and expert advice. The president of the Association for a Better Long Island, Jan Burman, wrote an editorial in *Newsday* attacking STAR and other activists who did not merely want to clean up the lab but to close it as returning "us all to the dark ages." "If the Brookhaven lab is closed," Burman concluded, "we will deserve the darkness that will follow."[17] Others viewed the prospect of shutting the HFBR, or the lab, as an "environmental Maginot line," a popular but useless political action that would seemingly improve public safety but would actually undermine it by providing a false sense of security.

The Montel Williams Show stimulated Leon Jaroff, the founder of *Discover* magazine, and so eminently regarded that he had an asteroid named after him. Jaroff had a reputation for exposing and ridiculing psychics and other pseudoscience. My "real hobby," he once declared, was "bashing the irrational." Jaroff found STAR's actions a perfect mark. Retired and living in East Hampton, he began writing to local newspapers, including *Dan's Papers* and the *East Hampton Star*, mocking wild and irresponsible claims by antinuclear activists and STAR'S East Hampton–based "rich and powerful glitterati." After Grossman won a journalism award, Jaroff blasted his misuse of facts and zeal to attack any sort of nuclear research. Jaroff wrote, "In all likelihood, he will soon be campaigning against the nuclear family."[18] As for Baldwin's show, "I think it is somewhat appropriate," Jaroff wrote, "that S.T.A.R. has now made its national debut on *The Montel Williams Show*, best known for such guests as two-headed prostitutes who are experts in needlepoint and pregnant women who play professional football."[19]

Jaroff would tweak Baldwin for years. When Baldwin called Jaroff "a whore for Brookhaven National Laboratory," Jaroff responded, "Unlike a

whore, I do not provide my services to the Brookhaven Lab for money. . . .
I defend Brookhaven against extremist know-nothings strictly out of conviction and out of knowledge that is obviously superior to yours." Jaroff then
noted that Baldwin, supposedly antinuclear, had appeared in the movie
The Hunt for Red October, which glorified the exploits of a nuclear-powered
submarine. Jaroff asked rhetorically why Baldwin had done so—then filled
in: "For the money, of course."[20]

One of the country's most distinguished health physicists wrote to the
New York Times about the show. Otto Raabe, a professor emeritus at the University of California, Davis, was president of the Health Physics Society, and
three years before had received that organization's Distinguished Scientific
Achievement Award. For two decades his organization had been struggling
to combat groundless fears about radiation. He was incensed not only that
celebrities were venturing into his field but also because their statements were
damaging as they prevented the public from having an accurate picture of
public health threats.

On January 22, Raabe wrote the *New York Times* a lengthy letter about
Caldicott and *The Montel Williams Show*. "She has been peddling these types
of irrational antinuclear falsehoods for about 20 years," he wrote, "under
the mantel of a medical degree from Australia, but she is not an expert in
radiation sciences and has no training or experience in the fields of radiation
biology, health physics, radiation risk assessment, or environmental health."
He described the Health Physics Society of which he was president. It is
a "national professional organization of over 6,400 scientists, physicians,
educators, engineers and operational health physicists who are dedicated to
insuring radiation safety for workers and the public. Many of our members
are professionally certified as experts in radiation safety by the American
Board of Health Physics. Thus—not surprisingly—we are distressed that
the media often lend credibility to the outlandish and unfounded statements made by Helen Caldicott." Her irrational and easily refuted remarks
about the supposed dangers posed by the lab are generating misleading fears.
"Where does she get this garbage? I have worked in the field of radiation
safety and environmental health for almost 40 years, and I can assure you
that these statements are totally false." He continued, "Unfortunately I can't

in this short letter specifically address all the stupid, outlandish, and ridiculous contentions and predictions presented by Helen Caldicott." Radiation exposures to people on Long Island, New York City and state, and all over the United States have been exhaustively studied by such eminent and trustworthy institutions as the National Academy of Sciences and the National Council on Radiation Protection on Measurements. The tiny amounts associated with Brookhaven National Laboratory are minuscule compared with natural background radiation and things like medical diagnostic X-ray exposures. "These background exposures," which are far, far greater than those that come from the lab, "are not hazardous!"

Raabe concluded, "I implore your readers to rely on recognized authorities and board-certified health physicists for accurate information about environmental radioactivity. These are persons who have studied, trained and worked to safeguard the public and the environment from exposure to ionizing radiation and who are truly the ones who are standing for truth about radiation."[21]

The *New York Times* did not publish Raabe's letter.

STAR's celebrities got on a nationally televised show and had standing in the media right down to society columns. Jaroff, an eminent science writer and expert in exposing health frauds, was able to publish a few articles in a small paper on the East End of Long Island, while Raabe, an eminent health physicist who was an expert in the relevant field, went unpublished. It was not that they had done anything wrong; Baldwin and Caldicott were only playing a game at which they, and not the scientists, were professionals.

The day after *The Montel Williams Show* was broadcast, the first DOE public comment meeting for the EIS was held at Longwood High School. John Axe, a scientist who worked at the HFBR, was heckled by a half-dozen or so activists sitting in the last row of seats against the auditorium's wall. At one point Axe mentioned the reactor's role in investigating a certain kind of cancer treatment. One activist interrupted in a loud voice, "Who did that ever help?" A person sitting in the second-to-last row, directly in front of the hecklers, turned around and said quietly, "Me." That silenced the activists, at least for a few minutes until their convictions returned. For scientists to have gotten more attention would have required such exchanges somehow to have taken place front and center onstage rather than in the back of the auditorium.

NO CANCER CLUSTER

The final report of the Suffolk County Task Force report, whose results had been publicized long before *The Montel Williams Show*—and whose members—included some STAR members—appeared two weeks afterward. "The Suffolk County Task Force on Brookhaven National Laboratory found no elevated cancer rates in communities near the Lab," it ran, and the incidences of rhabdomyosarcoma were found to be less in Suffolk County than in Nassau County, Queens, Brooklyn, and the rest of New York State.[22] Grimson told reporters, "There is no cluster of rhabdomyosarcoma in Suffolk County or in the lab," and nor was there a higher rate of breast cancer.[23] A follow-up study in 2011 by the ATSDR, the CDC's sister agency, reached the same conclusion: "The BNL site does not currently pose a health hazard."[24]

The cancer incidence issue had finally seemed laid to rest even before the Montel show. Not for the militant activists. Maniscalco called the Suffolk County Task Force report a "cover-up" and demanded that Grimson be fired.[25] Maniscalco pointed out that Grimson was employed by Stony Brook, now part of the lab's governance (though Stony Brook had not been when the study began and when most of its work was completed).[26] "All I can say is let the critics seek the same data and analyze it for themselves," Grimson told a reporter. "They will get the same results."[27]

STAR members continued to discount all studies of Suffolk County health hazards that did not link them to the lab, and kept demanding more studies that would, including studies of cancer rates in BNL employees. Since 1992, the DOE had been tracking the health of BNL employees, finding that "today's Brookhaven employees are, on average, healthier than the norm."[28] At the end of January 1998, responding to the pressure, the DOE announced that the New York State Cancer Registry would be compiling cancer rates of the 21,263 employees who have worked at BNL since 1947. The registry would then compare the rates of various kinds of cancers among these employees with those of Long Island and New York

State. The DOE expected the study to be completed in mid-1998. If BNL employees turned out to have a higher rate than the norm, the National Institute for Occupational Safety & Health would consider a full epidemiological study.[29]

STAR's lawyer Cullen called the Cancer Registry study "long overdue," but indicated that he would distrust the study if it resulted in another null result. "[O]ur concern is that they're going to come out with results that say there's no problem and that's the end of the story."[30] For STAR, the increased cancer just *had* to be there.

TRANSITION

After the diversion of the Montel show, the year 1998 looked promising for Brookhaven. Work on RHIC was proceeding on schedule. The DOE decided to allow the AGS to continue to pursue certain high-energy physics experiments along with its being the injector to RHIC. The environmental news was good. Suffolk County Director of Environmental Quality Joseph Baier, in letter to Bill Gunther, wrote that during 1996 and 1997 the SCDHS had sampled sixty-two public water supply wells on-site and off-site for tritium and found none close enough to the threshold to warrant investigation.[31] A ruptured pipe at the AGS leaked about 100 gallons of water containing slight amounts of tritium, but the Suffolk County Health Department said it was no danger. The two-month management transition from AUI to BSA was moving at a rapid clip.

D'Amato received positive news. At the beginning of his reelection year, the League of Conservation Voters announced that his environmental score had risen dramatically from the past two years (7 and 0, respectively, out of 100) to 29—"The Greening of D'Amato," wrote the *New York Times*—but it was still the lowest score of any senator in the Northeast.[32] After a year of Godfathering lab officials, and with BSA selected to replace AUI, the senator now professed friendliness to them, and on a lab visit put his arm on Bond's shoulder and said, "Great to see you, Pete! How are things going?"

The proposed cleanup budget news for the lab, however, was not rosy. The DOE budget proposal for the following year cut funding for environmental cleanup at BNL, but on February 9 D'Amato and Forbes requested an additional $11 million for cleanup at BNL, an action BNL supported. For years BNL had sought more cleanup money, but the DOE had responded that there were greater needs elsewhere. Still, after all the public uproar and DOE actions in 1997 the cut to the lab's cleanup budget baffled everyone—almost everyone. The *Long Island Voice* noted that Hanford's contamination was far greater, with a million gallons of high-level waste in the groundwater. The reporter wrote, "[I]n the coming battle over the budget, don't forget that we Americans have some bigger messes to clean up, too," quoting a Hanford source as saying, "A few curies of tritium wouldn't have even made my radar scope out here."[33]

In late January, Bond and Bebon made a last trip to DOE headquarters to brief Krebs, Lash, and Peter Brush (O'Toole's replacement) about the status of the many issues at the lab including funding issues that will face the new contractor in another month.

Toward the end of February a consortium of about thirty civic and activist groups called CALA, which included STAR, met with a group of BNL and DOE individuals, including Marburger and Helms. A previous request to have one or more of its members sit on the BSA board was described by Marburger as a "humungous" request and said that BSA had rejected it as requiring changes to its corporate structure. BNL representatives were puzzled when STAR's representatives asked about the long-studied Building 650 outfall, not realizing that STAR was planning a blindsiding involving that site for the next week. The CALA representatives also demanded the right to participate in determining the protocol for an employee health study and oversee its conduct. Helms agreed that input to the protocol was welcome but that nonscientific oversight would not be possible.

On February 27, AUI sponsored a brief celebration in Berkner Hall to thank Bond and Bebon for their efforts over the previous 10 months. Recalling Schwartz's words on the day AUI was terminated and Bebon became Deputy Director about the lab's future smooth sailing, Marge Lynch presented Bebon a glass boat. Lynch gave Bond a glass globe, declaring that he had carried the weight of the world on his shoulders. AUI also put a note in the *Brookhaven Bulletin* that appeared that day, marking the end of the organization's 50-year stewardship of the lab:

Together we have made many significant contributions to science, technology, and education. In addition, we have provided facilities to thousands of researchers who have enriched mankind with intellectual breakthroughs in virtually every scientific discipline. None of this could have happened without the skills and dedication of everyone on the Laboratory staff. Thank you, each of you— employee, visitor and friend—for making BNL what it is today, a world-class research institution. We are confident that BNL will continue to build on its outstanding record of scientific excellence, and we wish all of you every success in your future endeavors.[34]

The next morning, February 28, was AUI's last day as Brookhaven's manager and Bond's as interim director.

SCAMMED

The day began with STAR demonstrating that it was not going to give BSA a grace period; Rowe learned from reporters that it had scheduled a press conference on Monday morning about what was supposedly the lab's most lethal leak yet on the Building 650 sump. The lab's environmental health employees had been cleaning up the site for years, following a process outlined at public meetings, and the site had been discussed at several LIPC meetings. As Gunther put it, the Building 650 sump outfall "is one of the best-studied, most-understood pockets of contamination on the Laboratory site, and has been under the microscope of our Superfund process for several years."[35] But STAR was going to announce "new information" about the contamination provided by a supposed whistleblower named Robert Ramirez. STAR had not informed the lab, but had contacted the *New York Times*, which was running a front-page story on Sunday morning. Rowe obtained a copy of STAR's announcement from a reporter and passed it to Bond, who gave it to Marburger.

The term "whistleblower" implies someone who, at great personal and professional risk, exposes illegal, dangerous, or fraudulent activities from an organization with which they are associated—a Daniel Ellsberg, for instance, who was the military analyst who leaked the Pentagon Papers in the middle of the Vietnam War. The term was somewhat contrived when applied to Ramirez, a disgruntled employee with a discredited speculation. A former staff hydrologist who had worked in Brookhaven's Office of Environmental Restoration (but not in connection with the Building 650 area), Ramirez was found to have engaged in improper behavior in 1996, and the lab's human relations department gave him the choice of being fired or resigning without disciplinary action. Ramirez chose the latter. He eventually contacted STAR, telling them that he had a "theory" that the strontium-90 at Building 650 had mixed with other chemicals and flowed across the lab's boundaries, and STAR used the theory to stage a press event designed to disgrace the lab's new manager on its first day.

When Marburger met with STAR at their invitation that Saturday morning, its members had mentioned neither Ramirez nor the press conference to him. The following is from Marburger's diary:

> I went into this meeting skeptical that STAR was trying to lower my defenses in prep. for their press conference on Monday that Peter Bond had alerted me to. Had their notice to the press in my pocket but did not say anything about it until later in the meeting. Jim Tripp, Scott Cullen, [Carrie Clark], Peter Strugatz, David Friedson, Jan Schlichtmann were all there in the side room at Danford's when I arrived. Jan did most of the talking, saying STAR wanted to get away from the "terrorist" tactics and forge a new way of working with BNL that would be a paradigm nationally. He described his idea as setting up a respected independent 3rd party we could take our issues to for adjudication or neutral study. I said I liked the idea, that it wasn't a new idea, (I said it was a "no-brainer") *so* we have to ask why it didn't work before. I said I thought the reason was that STAR has wanted to be THE community rep, but I don't have the luxury of choosing; I have to deal with everybody, etc.
>
> At some point I revealed that I knew about the upcoming press conference and pointed out this also doesn't make it easy to work with STAR. Jan said, of course STAR had the ability to "zing" us whenever they want. He said they have the backing and the personnel to tie us up in knots at their will. But that's not getting anybody anywhere, hence today's meeting. I said I didn't appreciate it but since I have to deal with everybody, I wanted to focus on issues and the 3rd party mechanism is fine with me. So what next?
>
> . . . I said I probably would respond to the press conference by saying that I challenge STAR to find a new way of working together on issues & stop this press conference warfare.[36]

That evening the Bonds hosted a dinner for the Friends of Brookhaven and the Marburgers. Shanklin presented Bond with a "Friends of Brookhaven #0001" certificate; Shanklin joked that it should have been #007. During the party, Bond received a call from Rowe, who had learned from a good contact in the DOE press office—a career, not a political appointee—that STAR had told Secretary Peña about its press conference, and she confirmed that the *New York Times* was running a big story the next morning.

The usually unflappable Marburger slept badly that night. "The morning paper brought a disastrous story in the *NY Times*, totally without context and in my view irresponsible,"[37] so irresponsible that he asked a reporter to fax him a statement of professional responsibility for journalists that he could show to editors. The *Times* front page proclaimed: "Water Near L.I. Lab Reported in Jeopardy." A banner headline on the front page of the Metro section read, "At Brookhaven, a New Charge That Radiation Threatens Water,"

Figure 7.2
Attendees at a party at the Bond's the last night of AUI's management of Brookhaven to introduce Marburger to the Friends of Brookhaven. Left to right: Sean McCorkle, Jim Hurst, Ed Kaplan, Frank Marotta, Jean Jordan-Sweet, Jennifer O'Connor, Ben Ocko, Bill Graves, Joanna Fowler, John Shanklin, and Peter Bond. Courtesy Sandra Bond.

subheaded "Environmentalists Say a Site Could Harbor the Lab's Worst Pollution." The new development was presented as painting the lab more dangerous than ever before. Two days later, in a metro news brief that came not from its own reporting but from the Associated Press, the *Times* would reassess, quoting Ramirez speaking truthfully that "it's a theory."[38] But the *New York Times* reporter, caught up in STAR's hype, had investigated neither Ramirez's background nor his claims, reporting him as alerting Long Island to well-founded "fears" about Brookhaven's threats.

After Marburger saw the article Sunday morning—his first working day on the job—he called Bond and a few others to meet him in the director's office at 1:00 p.m. to handle media queries. Having slept poorly, for once he showed up to the lab in casual clothing—which for him meant that he didn't have a tie, only "my corduroy coat and a black mock turtleneck under a striped button-down shirt."[39]

There was nothing to the information, he and other BNL officials told reporters. The Building 650 issues were well known and under continuous surveillance by state and local environmental authorities for years. "My message to press," Marburger wrote, "was that the assertions of Ramirez were not new to us, that he had not in fact approached lab mgt. and been rebuffed, that his mechanism had been thought up by others, that testing subsequent to his departure from lab shows his theory is not operating here, & that there is no observed flow of radioactive contamination away from this site." Marburger continued, "It was an exhausting day. I regretted not being dressed properly to go on camera. Maybe I should keep clothing in my closet just for such occasions."[40]

When a National Public Radio reporter asked Gunther, who had studied the Building 650 area as much as anyone else, about Ramirez's claim that there was strontium-90 in wells near the south border of the lab, the usually reserved Gunther replied that there was zero data in support of that "bogus" assertion.

STAR's Monday morning press conference—billed as a "Meet the Whistleblower" event—was elaborately staged for the media. STAR headlined its press release "Undisclosed Radiation Threat Emanating from Brookhaven National Laboratory." "They don't have the evidence to disprove the theory," Ramirez said. "Let's find out the truth." STAR painted the event as a courageous whistleblower exposing deadly crimes by villainous government scientists. Few media stories—in the *New York Times*, *Newsday*, and other outlets—questioned STAR's interpretation. Some articles cited Suffolk County environmental official James Pim's remarks that Ramirez's claims were "very unlikely," and Gunther's that Ramirez's theory was "not enacted . . . not happening." But these careful remarks were not as exciting as STAR's apocalyptic ones: Cullen claimed that this new discovery was "worse than the other problems," and Caldicott that the lab was releasing "lethal soup."

The STAR members were upfront that they had scheduled the event on the day the new management would take over to show they were going "to hold the new management to a higher standard of accountability," as Friedson said.[41] Or, as Cullen put it, "If you don't play nice it's going to get nasty."[42] STAR had ensured that when Marburger gave his first open-house site tour of the lab that Monday, the principal topic of conversation was not science but strontium.

Jaroff was one of the few to find all of this amusing. STAR, he wrote in the *East Hampton Star*, has everything going for it: "wealthy backers, a covey of celebrities, a lawyer, a handful of pseudo-scientists, a full-time staff and (obviously) a number of willing, but naïve letter writers." I am unconnected with Brookhaven, Jaroff continued, "but I do know scientific illiteracy and falsehood when I see it." One of STAR's spokespersons had tried to account for the Suffolk County Task Force's finding that the East End's cancer rate was higher than around Brookhaven by claiming a "synergy" between Brookhaven's emissions, those of another reactor in Connecticut, and agricultural pesticides. Jaroff mocked, "Duuuh! We'll work Brookhaven into this anyway we can. . . . So there it is, STAR. You profess lofty goals, but your highly-visible spokespeople start with the assumption that Brookhaven is responsible for all of Long Island's ills, and then play loosely and irresponsibly with the Truth in dishonest attempts to prove their point." STAR would be performing a real service if it used its "substantial resources to hire a couple of genuine epidemiologists to investigate the South Fork's elevated breast cancer rate . . . instead of starting with the knee-jerk conviction that Brookhaven is responsible."[43]

Later that March, Baldwin and other STAR members held a $700-per-couple fundraiser in Manhattan at the home of Patti Kenner, a philanthropist who besides a Manhattan apartment also owned a condominium in Aspen, ranch in Colorado, and vacation home in East Hampton. The event was billed as "a very special evening to discuss concerns about chemical and radioactive contamination associated with Brookhaven National Laboratory." One speaker likened the lab to "Hitler cooking babies." Another said that the DOE's water hookups "prove" the guilt of the lab. More plumes were being discovered every day, attendees were told, and claimed that a lab spokesperson had asserted that "people must be sacrificed for the public good."[44]

CULTURE CHANGE

Martha Krebs had told one interviewer that the Brookhaven decision was the "next step" in a culture change at DOE that had begun half a dozen or so years previously, following the Tiger Teams, to elevate and integrate safety into the management structure of all the labs.[45] Ever since the Tiger Team visits of the early 1990s the DOE had been agonizing over how to change lab culture, and the tritium leak—and O'Toole's impetus—had finally provided the opportunity that prodded the agency into action.[46]

Many lab employees didn't like it, for under BSA the culture change was a likely turn away from the basic research they had assumed was their mission and that they were used to. BSA Board member Bill Madia all but promised this turn when he told a *Newsday* reporter, "You will have certain members of the basic science staff who just want to do basic science, but we will work very hard to convert basic science to applications to see tangible results." Worse, Marburger seemed to go along, telling the reporter, "The idea is to make BNL a little more like Pacific Northwest, a little more businesslike in the way it operates."[47]

Lab operations under BSA management almost immediately became more troublingly bureaucratic than under AUI. Under AUI, management had been effectively bottom-up, and focused primarily on the scientific and programmatic progress. Under BSA, Brookhaven's management became more top-down, with programmatic decisions largely in the hands not of department chairs but of a management council, the majority of whose members were nonscientists. The new contract required department chairs having to focus on goals not initiated by their scientists, but handed down from the laboratory administration—and the department chairs would then be evaluated against those goals.

The new management thus indeed brought a culture change, as DOE wanted but as scientists feared. Scientists found working at Brookhaven more and more like working for business than a university. Employees who had arrived in AUI days soon chafed at such things as having to use time clocks not in a pro forma way, with one's managers trusting that your time was appropriately spent, but as a carefully itemized and scrutinized ledger of your activities. It became no longer possible to regard the laboratory community as a big family whose members naturally helped each other out, and from which one might retire, but as a corporation, working for which should be regarded as a career step. "I was hired by AUI as a scientist," one senior staff member said a few years after the transition, "and BSA taught me how to be an employee."

Much of these changes came about from a combination of the growing size and complexity of the federal bureaucracy and of instruments such as the Relativistic Heavy Ion Collider and the National Synchrotron Light Source II—the successor to NSLS I. Still, the new management exerted a strong force in creating a bureaucratic climate far from the well-equipped scientific university atmosphere envisioned at the beginning of GOCO contracting. Formal paperwork was sometimes filed where it's hard to tell whether it was required or a satire of the requirement, as when an "Adverse Event Report" was drawn up after a participant in a clinical trial got a bee sting.[48]

It wasn't just at Brookhaven, though; the DOE was changing lab culture everywhere. After the Brookhaven competition, the DOE's competitions for other lab contracts increased and were held at some places that had never been re-contracted. An even greater impact was that new contractors were expected to include at least one large corporation—a Battelle, a Westinghouse, a Lockheed—to handle the operational aspects of the laboratory, including environmental cleanup, which had close ties to the government and which was more inclined to regard an academic-like atmosphere with suspicion. DOE contracts became more prescriptive and aligned laboratory operations more with federal practices.[49] "It's been my experience that Battelle does whatever the Energy Department tells them to," said a Hanford employee.[50]

The DOE now looked for contractors to operate national laboratories that were a combination of science-based and operations-focused organizations—but this dual character could, and sometimes did, result in problems, as their priorities were not always in sync.

In 2001 DOE gave a grant to the Georgia Tech School of Public Policy to report on how three recent new contracts—at Brookhaven, Oak Ridge, and Sandia—had impacted the labs.[51] The study of Brookhaven revealed that the change created significant "disruption, anxiety and uncertainty" at the lab, which was not surprising given that Peña had terminated the contract with no advance warning and without any period of evaluation or review. A few managers regarded the changes implemented by BSA as simply implementing modern management techniques, while others lamented DOE's new control and influence over the lab. "Some scientists, however, found the procedures silly and cumbersome. One example of the sources of their concern came in the form of a requirement to state in their program goals how many important scientific discoveries they would make in the next year. Some employees in the lab found this requirement so ill-conceived as to be amusing, and they posted this requirement on their office doors and bulletin boards for humorous effect."

The report also found contracting for management of the laboratories did not result in "market competition," pointing out that, at BNL, only two bids were received, each of which was a mad scramble. Bidding is expensive, and only a few corporations have the experience, technical breadth, ability to assume liability, or opportunity to profit from tech transfer to make it worth spending millions of dollars to mount a bid.

NEWSDAY AT BNL

In the spring of 1998, *Newsday*'s managing editor Howard Schneider proposed that the newspaper "embed" a team of reporters at the lab, giving them freedom to stay as long as they wanted to write a series of in-depth stories. Marburger, an advocate of deep transparency—and knowing that once the reporters investigated the inner workings of the lab they would find no deliberate cover-up and no serious contamination—agreed, and accompanied Rowe to a meeting at *Newsday* to work out details. Marburger ensured that the *Newsday* reporters had gate passes, an office, the freedom to visit all facilities and speak with lab employees, and access to all reports, maps, data, and other documents they wanted.

It was the same arrangement that the lab had made with the CWG two years before—and it, too, was a risk, both for Brookhaven and for *Newsday*. For the lab it meant the possibility that the reporters would use the access principally to find evidence to back up sensational negative coverage. For the newspaper it meant taking highly paid first-rate investigative journalists off the beat for more than a year, and finding little of interest to readers. Schneider also knew the reporters inevitably would be viewed as partisan: at least some lab employees would view the reporters as scandal hunters if they found something, while STAR members would regard the reporters as apologists if they didn't. "But the lab was so controversial that we saw it as crucial," Schneider recalled. "As the region's leading local newspaper we felt it was our mission to get to the bottom of the story, to provide reliable information and context. And it came at a time when *Newsday* was financially successful, had enough financial resources, and had the right reporters."[52]

The lead and principal reporter was Charles V. Zehren, assisted by photojournalist John Paraskevas and reporters Jordan Rau, Jack Sirica, Lauren Terrazano, and a few others. "Charlie had just come back from being an investigative reporter at the Washington desk," recalled Rau. "He had a 'We have to get to the bottom of this' attitude." Zehren was indeed diligent. He interviewed everyone from groundskeepers to Marburger. He went through environmental and radiological training, and was given his own dosimeter so that he had access to all facilities including ones doing nuclear research such as the still-shut HFBR, the decommissioned Brookhaven Graphite Research Reactor, and the "hot labs" where radioactive material was studied. Zehren inspected all the accelerators, including the Alternating Gradient Synchrotron and the yet-to-be-completed Relativistic Heavy Ion Collider. He visited the PET lab—where Nora Volkow enthusiastically offered to scan his brain, though Zehren refused. He visited the Protein Data Center and the animal and plant testing facilities. He scrutinized the administrative structure and practices of the lab, and its arts and cultural programs.

Zehren also pored over all the available documentation, including Occurrence Reports sent to the DOE, and he did not turn up anything unreported of major consequence.

The first article in the five-part series called "City of Science Under Siege" was published in May 1998; four other articles appeared in November.[53] The series, which examined the science, culture, community relations, and environmental issues at the lab, showed careful and comprehensive coverage about a complex institution operating in a challenging environment. The *Newsday* editors nominated the series for a Pulitzer Prize. In the end, *Newsday* cited sources that claimed the pollution emanating from the lab posed health risks as well as others who said it did not, including some local, state, and national groups who said that while there were some past disposal practices (some not from the lab itself) that would now be unacceptable and needed to be cleaned up, neither the chemicals nor radiation at Brookhaven posed a health threat to employees or the community. The journalists indeed did not find a Big Story in the sense of environmental health impacts as at Love Canal, Woburn, Toms River, or Hinkley after all. A radio interviewer asked Zehren and Sirica specifically what they had discovered about the lab's public health threat posed by either radiation or chemicals. "I must say that at the end of this you cannot document anything resembling a huge public health crisis here," Sirica replied. "We could only come up with 4 or 5 houses where *possibly* you had a well that got in that pollution."

The Big Story, it turned out, was not about environmental danger. The Big Story was entirely different—about the reasons why a small leak of tritium of no health hazard had brought about such a huge impact on the lab and the national laboratory system, and what it revealed about American public and political attitudes toward science and scientists.

GRACE OF GOD

In July, the Department of Energy held an "Office of Energy Research Senior Managers Forum" to focus on DOE's interactions with its labs, including a discussion of "Lessons Learned" from the BNL events the previous year. Martha Krebs, Tara O'Toole, Peter Bond, and Mike Holland of the DOE's on-site office made presentations, as did Crease, Gary Boss of the GAO, and Judy Jackson from the Fermi National Accelerator Laboratory.

"There but for the grace of God" was Judy Jackson's first slide. If these events could happen at Brookhaven, she said, they could also happen at any major US scientific facility—even Fermilab, which had no reactors.

Krebs pointed to split responsibilities between science, operations, and ES&H both within DOE and in the labs. The disconnect with Congress was another problem: the labs think primarily about science and Congress about budgets. Congressmen, too, need to learn to think in a broadly strategic way about science, she said, and to view DOE laboratory research as an entire system rather than just a collection of programs.

O'Toole found the Brookhaven events an "organizational accident" so complex that the usual corrective steps—more standards or regulations—would not help. Her list of triggers included the spent fuel pool issue, community concerns, changes in AUI and DOE leadership, political opportunism, the media's penchant to assign blame, and the DOE's tendency to play "Gotcha!"

Boss was harshly critical of the seriousness and depth of DOE oversight, saying that Peña was unaware of DOE's "excellent" rating of AUI when he terminated its contract, evidently not having carefully read the Options Document. Mike Holland of the DOE's local BNL office reported on the corrosive atmosphere in which "urgency was associated with each and every event and there were twice a day calls to headquarters in DC."

Bond noted that, in 1997, the DOE heaped criticism on BNL's environmental health and safety performance and public outreach, yet these had played only a minor role in the way the DOE established the amount of the contractor fee. The DOE had also not provided timely funding to fully decommission the BGRR, which resulted in water in its air ducts to accumulate three decades after its shutdown. The DOE often needlessly micromanaged BNL, for example, insisting that it approve every press release even about trivial things like music concerts. DOE headquarters sometimes did not read, shelved, or ignored important memos; an example was Attorney General Vacco's request for a summary of the meeting in his office, which had gotten "lost" in the DOE's Washington office and was delivered over half a year later. The DOE had a propensity to take credit for the work of others, as it had for Schwartz's ES&H plan after firing him. On the other hand, Wagoner and Helms had shown there could be a greatly improved and necessary relationship between BNL and the DOE, and that kind of partnership needed to continue. Bond did not spare AUI and the lab, saying that BNL had to place a higher priority on ES&H matters and establish better relations with local Long Island officials and the community. At least part of the cause of these problems was not changing leadership in a timely manner.

Crease said that the deep background to the BNL events lay in five overlapping factors: the changing public enthusiasm for big science after World War II, the US government's changeover to a different administrative structure for managing the national laboratories, a set of regional factors having to do with Brookhaven's location, increased environmental sensitivities, and poor understanding of how to engage the public about scientific research.

At one point in the conference, a scientist (the authors do not remember who) outlined his carefully worked out approach for handling controversies such as Brookhaven, and concluded by saying, "It's the perfect solution, but it's not implementable." The room was suddenly silent. Then, from the back, a voice articulated softly and clearly what was certainly the most important lesson learned: "If it's not implementable, it's not a solution."

Plenty of "lessons learned" emerged from the conference that were clearly crucial to the health of US science. Could they be implemented? DOE officials were optimistic about their ability to change deeply embedded habits in the agency, but other attendees were not so sure, remembering O'Toole's own fondness for "Gotcha!" Other problems went unmentioned, such as the extent to which the DOE had fired AUI as a lesson to other labs rather than as a measure directly addressed to conditions at the lab.

Also unmentioned was the still-ongoing EIS process that the DOE had assured everyone would determine the HFBR's fate. It would be a test of the DOE's resolve to learn the lessons.

Judy Jackson had expressed it well: many US scientific facilities were surviving on luck alone. For the HFBR, it was about to run out.

8 SCRAMMED

It was a political decision, but it doesn't matter. The important thing is, it's closed.

—BILL SMITH

NEW DOE SECRETARY

In April 1998, Peña announced he was leaving the administration, barely a year after becoming DOE secretary. (Ironically, one of his last challenges on the job was finding a way to make tritium for the defense industry.[1]) To replace Peña, President Bill Clinton picked the US Ambassador to the United Nations Bill Richardson. Early that summer, Peña briefed Richardson on half a dozen or so key decisions that he would have to make. The decision about the fate of the HFBR was one, and it had created a buzz. Even before being confirmed in July Richardson was heavily lobbied, inundated by letters from activists and by impassioned phone calls from D'Amato and Forbes, who were both facing reelection.

Before becoming UN Ambassador, Richardson had been a congressman in New Mexico's third congressional district, which included the Los Alamos National Laboratory. Still, Richardson was a canny politician, known to befriend a wide variety of constituents, and he was not known as a particular advocate of science. Furthermore, while UN Ambassador he had made frequent trips to the Hamptons, participating in celebrity events such as softball games (authors vs. politicians) along with Baldwin and others, at which STAR members began lobbying him.[2] One of the most insistent was a new STAR recruit, supermodel Christie Brinkley.

"I became involved because I was a mom," Brinkley recalled. At the time—the end of 1997—she was living in the Hamptons, had two children, and was pregnant with a third. Having grown up in California where she had played happily outdoors on its beaches and hillsides, she wanted the same freedom for her kids. One day she was galvanized by a talk she heard Caldicott give about the nearby power reactors in Connecticut, New Jersey, and New York. "I was really taken aback, because I had *no idea*. I thought that I was out in the fresh sea air of the Hamptons," she said. "Then when I found out about the Brookhaven High Flux Beam Reactor, and the fact that it had been leaking radioactive tritium unnoticed for twelve years, I was *horrified*. And I was alarmed by the cancer clusters in Brookhaven, and about the cases of rhabdomyosarcoma, which were in a cluster in that area. Then when you put that together with the fact that the High Flux Beam Reactor wasn't doing research that was benefiting humanity—they were working on things like electronic ways to raise and lower windows in your car and stuff like that—I thought, 'This has got to close.'"

When Alec Baldwin called up Brinkley to recruit her into STAR she was very willing. She would soon prove to be STAR's avenue to the new DOE secretary.

STOPPING THE RESTART

Few scientists at the DOE were worried about restarting the HFBR, and continued to incorporate the facility, along with its upgrade, into their future plans. "It was the finest high beam neutron source anywhere," said William D. Magwood IV, who in June of 1998 had replaced Terry Lash as director of the Office of Nuclear Energy, Science and Technology. "I don't know anyone who took exception to that description. From the technical standpoint it was very clear that the reactor was safe, very well operated, in excellent condition, and that the neutron science community wanted it very badly. I don't think that it was in anybody's imagination that the leak could lead to the reactor being shut down."[3]

At the beginning of 1998, the DOE had planned to release the draft EIS in August 1998 and the final version in November, with Peña saying that he would make the final decision in December 1998. Members of the scientific community anticipated an eventual positive outcome, and prompt completion would have made a restart easier. Activists, however, knew that delaying an EIS causes huge cost overruns and often kills projects, and they forced a series of delays. These hurt the chances of a restart for three reasons. First, it gave STAR more time to mobilize opposition to the reactor. Second, the delay forced neutron scientists to look, plan, and move elsewhere to do their work—and it gave others whose facilities competed for money with HFBR within DOE's Office of Energy Research budget time to mobilize, including supporters of the proposed Spallation Neutron Source (SNS) at Oak Ridge. Third, the delays became a reason in itself. The longer the process took, the more persuasive grew the argument that it would become so protracted and eat up so much money that it would be better to invest resources into operations or new facilities more likely to bear scientific fruit, eventually leading some DOE officials, including Patricia Dehmer and Martha Krebs, to have second thoughts about the restart. Each year, they were spending over $20 million on the HFBR, for it still required a full staff to keep it on standby. Preventing the delays would have required strong leadership at the DOE. An additional block to any restart was the budget amendment from D'Amato and Forbes that prevented any funds being spent on a restart; even after D'Amato left office, Forbes continued inserting the language into the federal budget documents.

The DOE's attention was beginning to focus on other projects. In the early 1990s, as part of a deal to distribute several major projects to different laboratories (among them RHIC for Brookhaven), Oak Ridge National Laboratory was promised a new, more intense reactor, the Advanced Neutron Source (ANS), which would provide welcome relief for the heavily used HFBR. But for several reasons, including cost and the DOE's aversion to the prospect of promoting a reactor project in the post-Chernobyl era, the project languished and was terminated early in 1995. Dismayed, the neutron community sought another idea to keep hopes alive for an intense neutron source that was not a reactor. One promising possibility was an accelerator-based neutron source, which would fire protons into a target material such as mercury, shattering its nuclei to produce neutrons. Such a facility, the

Intense Pulsed Neutron Source (IPNS), had been built at Argonne and proven its value, and in 1997 the DOE began planning an even bigger facility, the Spallation Neutron Source, as a substitute for the ANS at Oak Ridge, raising the possibility of competition for resources with the HFBR.

Meanwhile, phony accusations, political posturing, and lab missteps persisted for months. Pete Maniscalco continued to claim that the out-of-operation HFBR was in danger of a core meltdown, despite the fact that all its fuel rods had been removed and shipped offsite.[4] Bill Smith claimed that the lab's "toxic and radiological agency" was responsible for the death of fish, turtles, and snakes in a Peconic pond three and a half miles downstream, but the pond's chemicals proved well within New York State limits, and no dead turtles or snakes could be found.[5] New York State had detected minute amounts of radioactive material in sediments in 1996, including americium. With the new cooperative testing agreement (BNL, Suffolk County, and Fish Unlimited) in place, new samples were taken on-site and off-site. In June the lab reported detection of plutonium in Peconic River sediment. When the data were examined more carefully, it appeared the finding may have been premature (demonstrating the need to recheck data before releasing it), but there was back and forth for several weeks as to whether it was valid and, if so, how much was from the US military's atmospheric nuclear bomb testing in the 1950s (the source of other radioactive contamination in the ground) and how much might have come from the BGRR. Finally, to stop the bickering, Marburger stated that if the level of plutonium was such that, if it needed to be cleaned up, BNL would accept responsibility.[6] He added that the episode illustrated the danger of demanding that the lab make public "every atom" of radioactive material it finds. "It creates a sense of panic that's just not justified."[7]

In October 1998 Richardson stopped at Brookhaven and addressed the recently-formed Community Advisory Committee (CAC), announcing that he would make the HFBR decision the following June, meaning another six-month delay. Then he visited Baldwin and STAR members in the Hamptons. Cullen attended the CAC meeting and told Grossman, "The language and message I got from Richardson is that they are going to restart this thing." Despite sign-carrying anti-HFBR protesters, Cullen reported that Richardson said his sense was that the majority of the Long Island public does not have a problem with restart, and that Richardson felt that the reactor was safe—running contrary to STAR's entire case—and Cullen came away with the impression that the DOE would do a "snow job on the public" and restart the reactor.[8] At a subsequent party in the Hamptons, Brinkley—one of several high-profile fund-raisers for the Democratic Party—tracked down Richardson to make a brief introduction of herself and the issue she was promoting.

STAR redoubled its efforts, armed with the knowledge of Richardson's ambitions to become the vice-presidential candidate behind Al Gore in the next election. On January 13, 1999, Richardson met with STAR, appeared to anoint the organization as the liaison between the DOE and the community, and gave STAR the go-ahead to do its own safety review of an HFBR restart.

Hearing of the meeting, Friends of Brookhaven sought to meet with Richardson as well. While their group lacked "notoriety, funding or media coverage," and its membership was "made up of [neither] wealthy Hamptonites nor actors," still, they consisted of "several hundred people who see great value in responsible scientific research."[9] Richardson did not respond. *Science and Government Report* was incredulous about the meeting between STAR and Richardson. "A rabidly antinuclear group has gotten the okay from Energy Secretary Bill Richardson to conduct its own safety review into the possible restart of an idled research reactor on Long Island, NY," it wrote. "What's more, the Department of Energy could end up paying for a study whose outcome is preordained to come out squarely against restart." Citing Cullen's remark that STAR's review will "have no impact" on Richardson's decision, the article continued, "How the politically suave and diplomatically inclined secretary could ignore a report reflecting a vocal, however emotional, segment of the area's residents is almost as puzzling as the question of how an organization whose website is named 'noradiation.org' and proclaims 'there is no such thing as a safely operating nuclear reactor' could possibly prepare an objective review on any reactor."[10]

What's New ridiculed the veep-wannabe. "To appease the tycoons and celebrities who have homes in the neighboring Hamptons, Secretary of Energy Bill Richardson decided to let the group Standing for Truth Against [*sic*] Radiation, whose spokesman is movie star Alec Baldwin, conduct its own safety review of the idled High Flux Beam Reactor. Scientists who have dealt with STAR think it's a really bad idea."[11]

In an appearance of balance, the DOE asked the Nuclear Regulatory Commission to conduct an assessment of the HFBR. "Actions taken to characterize and control the tritium plume were conservative, and this plume does not represent a radiological hazard to public health or safety," the NRC found after a six-week on-site survey, saying that it found "no safety significant issues" in restarting the HFBR.[12] "Uh, does it mean that the decision to shut down HFBR and terminate the Associated Universities contract was premature?" mocked Robert L. Park. "Of course, we won't know for sure until the STAR panel completes its safety review."[13]

NAVIGATING BY STARLIGHT

The EIS delays were stretching out and out. In his initial announcement in 1997, Peña had confidently declared that the decision about the HFBR would depend on and follow directly after the EIS, to be completed that fall with a decision in January 1998. Late in 1997—after the Forbes-D'Amato press conference announcing their introduction of an amendment to prohibit money for a restart—Peña extended completion of the EIS until November 1998—after the election—with the restart decision to follow. Then he announced that he was leaving office in April. On Richardson's visit to the CAC in October 1998, he promised that the decision would be made in June 1999. In April 1999 the draft EIS draft concluded that "the environment and public health and safety would be protected under any of the four reasonable alternatives for the future of the HFBR." STAR members, including Baldwin and Brinkley, met with Richardson again, after which he kicked the can once more, extending the comment period for another ninety days. Meanwhile, Krebs and Pat Dehmer, whose office funded the operation of the HFBR, told Richardson that the BESAC conditions—timely and cost-efficient restart, ability to get to 60 megawatts—would be unlikely before 2002.

On learning of this meeting with an actor and a supermodel, lab scientist Julius Hastings was appalled. Hastings called Richardson's office and asked for an appointment. He was refused. While Richardson had met with actor Baldwin and supermodel Brinkley, he not only turned down Hastings, an eminent scientist and fellow of the American Physical Society who had helped design the HFBR and had worked there since its beginning, but members of his office also reprimanded Hastings for seeking a meeting. "Secretary Richardson looks to the stars for guidance," Park remarked in *What's New* for the second time.[14] Hastings's action was appreciated at the lab; Deputy Director Thomas Sheridan gave him a bottle of fine whiskey.

Meanwhile, the HFBR remained on the DOE's list of key neutron facilities for the US program. Repairs to the reactor pressed on, and the approval process for the upgrade to 60 megawatts continued. But the DOE's bureaucracy was effectively paralyzed as it agonized about how best to proceed, dreading the inevitable controversy that would follow completion of the process.

In April, Richardson agreed to meet with Robert Birgeneau and Frank Bates, a chemical engineer from the University of Minnesota, who had been chair of the steering committee for the Advanced Neutron Source at Oak Ridge, which had been replaced by the Spallation Neutron Source. "The HFBR was the best neutron beam facility in the US," Bates said, "and losing it was a real tragedy. We said then that it would never be replaced, and it still hasn't."[15] The meeting did not go well. "Richardson was aloof, sat at the periphery of the group we met with and did not say much at all," Bates recalled. "Our pull with the Clinton Administration couldn't match that of the Hollywood crowd."

In October, Birgeneau was informed that he was being awarded the 2000 J. E. Lilienfeld Prize, given out by the American Physical Society. This international award, whose previous winner was Steven Hawking, recognized "outstanding contributions to physics by an individual who also has exceptional skills in lecturing to diverse audiences." Birgeneau promptly wrote Richardson a letter in which he noted that his contributions had included research at the HFBR on high-temperature superconductivity, in which he knew Richardson was interested, and reminded the secretary that the HFBR's shutdown had seriously hampered that research.[16]

"Bill," Birgeneau concluded, "I would love to be able to say in my Lilienfeld Lecture 14577 next May 1 that Secretary Richardson has authorized the restart of the HFBR and that our research can now proceed again in full force." However, on October 21, 1999, STAR arranged yet another last-minute lobbying meeting. Christie Brinkley, her husband Peter Cook, and Scott Cullen, who were in Washington, met with Richardson to pressure him to close the reactor. Inside the Forrestal Building excited whispers flew that "Christie Brinkley is in the building!" Cullen said that STAR had commissioned a poll that had shown "overwhelming public opposition to opening the reactor." He also told Richardson that STAR would take legal steps to block a restart.[17] At one point Richardson asked whether the HFBR restart would be "another Shoreham." Brinkley said, "No," then paused. Staring at him, she said, "It would be worse."[18] According to a report in *George* magazine, a then-popular political periodical, Brinkley "reminded Richardson that his aspiration to be Al Gore's running mate [in the 2000 presidential election]—a job he hadn't been coy about lobbying for—would be seriously compromised if he didn't acquiesce." In response, according to the article—which was clearly based on lengthy interviews with several participants—"Richardson's jaw dropped." The article continued, "Richardson didn't savor the idea of going up against a group supported by New York's media elite and spearheaded by a bunch of local celebrities."[19] Richardson asked his top aides whether he had to wait for the EIS to be released, and was told no. Richardson then told Brinkley that he was going on a trip, but when he got back he would finalize the matter.

"Secretary Richardson looks to the stars for guidance," Park wrote again in *What's New*.[20]

BLINDSIDED

"I'm a politician—I listen to people," recalled Richardson about his decision.[21] He listened to the local political leaders, most notably Forbes, now a

Democrat. D'Amato had meanwhile lost his reelection campaign, which did not include any substantive issues related to BNL, but the Forbes-D'Amato amendments to deny funding to an HFBR restart certainly were an important factor. He listened to STAR, whose members included Democratic Party influencers; he met with them several times, but not lab scientists or administrators. When asked about the parallels to today, when a minority of people have mounted a campaign against things like covid vaccines and climate-change measures in the face of scientific consensus, Richardson said that it was different; DOE scientists were divided, with scientists in the nuclear office pro-restart, while others, including Dehmer and Krebs, were leaning against despite knowing that no technical or environmental issues stood in the way of restarting the reactor. Only the scientific community was strongly in support of the HFBR's restart; all the other people Richardson listened to were at best lukewarm.

On November 15, twenty-five days after the meeting with STAR, Secretary Richardson decided to terminate the HFBR, aborting the decision process that the DOE had been planning for two years. Richardson's press release stated the termination was for "programmatic" reasons, not for any safety or environmental issues. The reactor was permanently shut down, or "scrammed," in the lingo of reactor engineers. Richardson did not consult with or inform anyone at Brookhaven and did not have a staff person inform anyone at Brookhaven, despite the severe impact that this decision would have on the lab. He did not consult any members of the neutron physics community, though he had met with Bates and Birgeneau. Nor did he inform anyone at DOE's local Brookhaven office. He told activists, reporters, and politicians, but left the HFBR's biggest stakeholders—its administrators, operators, and users—out of the loop.

This was the fourth blindsiding of the lab in two and a half years. In May 1997, the DOE had terminated AUI's contract without consulting either the AUI or the lab; in September 1997 Forbes and D'Amato had introduced legislation to terminate the HFBR without informing the lab; in March 1998 STAR had staged a press conference on the day of the new management takeover without informing the lab; and in November 1999 Richardson had terminated the reactor without warning.

Rowe was the first at the lab to hear.[22] Arriving home at 8:15 that evening, following a musical event, she found several urgent messages on her answering machine from *Newsday* reporters. They informed her that Forbes had tipped them off that Richardson had announced his decision to terminate the HFBR; *Newsday*, the reporters said, was going to run a story the next morning based on what Forbes told them and could not wait for a comment from anyone at Brookhaven—they would be running the story with or without a remark from the laboratory. Rowe then received calls from reporters from the Associated Press, whom Forbes had also alerted, and the *New York Times*. The *Times* reporter was annoyed that Forbes had "blown" its own extensive story about the lab with something that scooped it.

When she called the reporters back, Rowe said that she could not say anything because nobody from the Department of Energy or Congress had notified the lab, so she could neither confirm nor deny the news. The only thing she could say was that, if the news were true, it came as a surprise to the lab given that a decision process had been in place for two years that had been predicated on the conclusion of the EIS.

Rowe knew that Marburger was away in New Mexico, at dinner with Bill Madia, the director-designate of Oak Ridge National Laboratory, before he was scheduled to give a keynote address at the DOE's annual pollution prevention conference in Albuquerque the next morning. Rowe therefore called Acting Director Thomas Sheridan, who stood in when Marburger was away; he had not heard the news. Neither did the next person on Rowe's call-down list, Steve Shapiro, an HFBR experimenter who was good at explaining the reactor's work to reporters. When Rowe called Michael Holland of the DOE local office, Holland had only just learned of Richardson's action from the DOE's Chicago office. Rowe called Lillian Kouchinsky, Marburger's secretary, to try to reach Marburger. Kouchinsky said that Marburger was likely to arrive back at his hotel in Albuquerque at 1 a.m. New York time. Rowe called the hotel and left a detailed message.

When Marburger returned to his hotel room in Albuquerque Monday night, he found the message light flashing. One call was from Rowe, informing him of what the reporters said and that they were seeking a comment if possible. Another call was from Marge Lynch, who was visiting

Argonne for a meeting of her communications counterparts at other labs, and had seen a draft announcement by the DOE that Richardson was terminating the reactor. Marburger was shocked by its contents. "I consulted with scientists, the community, members of Congress and other elected officials," Richardson's announcement said. Restarting the HFBR would be costly and take until at least 2002, and the secretary wanted to devote resources to facilities that did what he called "productive research." The draft announcement said nothing about the fact that both the NRC report and the (already completed) EIS had concluded that restarting the HFBR would be safe. Marburger typed a statement for the press, emailed it to Rowe, and tried unsuccessfully to sleep. But the lab couldn't use Marburger's statement unless it had the DOE's approval.

Early the next morning, November 16, Rowe called Jeff Sherwood, her counterpart at DOE headquarters, to confirm the news. Sherwood said that the DOE had written up a draft press release but it that it had not been finalized, one reason being that they had written in a quote for Marburger and had not reached him to get his approval.

In Los Alamos, Marburger checked out of his hotel early and traveled to Albuquerque to deliver the keynote address. Richardson was supposed to speak but did not show up; Richardson's Deputy Secretary T. J. Glauthier did, but seemed uninformed about the decision. That evening, Marburger took a red-eye plane back to Long Island, arriving Wednesday morning— "too late," he thought, "to share the first wave of distress with others who had hoped for a better outcome." He called reporters from an airport phone booth. The reporters, he thought, sounded unusually sympathetic, and he asked one why that was. "Because the Lab is now the underdog," one told him. (The decision was "ill-considered," editorialized *Newsday* on November 17, and the next day "Know-Nothings Win the Reactor Dispute."[23]) On the morning of November 18, Marburger appeared on a 7:00 a.m. radio show on WLUX, hosted by *Newsday* reporter Ed Lowe. Marburger asked Lowe what he knew of the supposed dangers of the HFBR, and Lowe admitted that he had only "skimmed" the reports.

Sheridan made a blistering statement to the press, and was called to Washington to be severely reprimanded. Outrage about the decision, and the DOE's failing either to consult or to tell the lab rippled through the lab, the CAC, the US neutron community, and the science community generally.

POLITICAL DECISION

At Brookhaven, a routine meeting of the Community Advisory Committee (CAC), formed the previous year, had been scheduled for Thursday evening. The CAC was one of BSA's very important and positive contributions to community relations. A moderated independent group that met, and still meets, monthly, it brought the lab and community closer together. The CAC, which consists of a variety of community members representing many civic and community organizations—including some of the lab's most vociferous critics—has minimized the impact of extreme groups and sensational headlines.

This particular CAC meeting was, in a sense, the culminating moment of all the events in the nearly three years since Paquette's discovery of the leak. Ever since, the DOE had justified and defended its actions involving the lab—including its termination of AUI's contract, its launching of an (unnecessary) Environmental Impact Statement with its public participation component, and ultimately its shutdown of the reactor—based on what the Department of Energy's officials claimed was the community's loss of trust in the lab. The CAC had been set up with a full spectrum of the Suffolk County community—its civic associations, community groups, interested individuals, and local activists—to engage with lab scientists and staff to achieve just that. Lab director Marburger was an active presence at the meetings, and had given CAC members the feeling that they were co-participants in lab activities. The DOE had firmly promised CAC members its cooperation and participation in all major decisions affecting the lab, the community, and the environment.

The DOE had now betrayed them, in its most important decision so far about those three things. For the DOE had promised that it would wait to make the decision on the HFBR restart until after the EIS had been completed. The CAC members had just learned that the DOE had already acted; it had terminated the HFBR without their input or knowledge and without, as promised, awaiting the outcome of the EIS. The agency had done so after listening not to their voices but to those of Democratic Party influencers. The CAC members—even many who were opposed to the restart—were furious.

On Thursday morning, the DOE caught wind of the rage about to erupt at the CAC meeting about the betrayal. Wanting its decision to appear careful, reasoned, and professional, it ordered Moniz, Krebs, and Dehmer to the meeting. Behind the scenes, the action was chaos. Moniz made it, but Krebs couldn't and appeared on speakerphone. Dehmer happened to be visiting Argonne National Laboratory outside Chicago that morning. "I was 'found' at about 11:30 a.m. at ANL and told to be at BNL by 6:30 p.m.," she wrote in a subsequent email. "No time to change clothes or do anything but hop in the rental car and take off. I made it to Berkner Hall with about 10 minutes to spare!"

All parties were represented at the CAC meeting. Brookhaven administrators and scientists were present, among them Marburger, Shanklin, and Rowe. Local on-site DOE representatives were there, including Crescenzo. The entire range of taxpayers' associations and groundwater groups were also present—including Mannhaupt, Pannullo, Esposito, and others who had been involved since long before the leak—and a few STAR members. But two-thirds of the thirty-nine CAC members were said to be intending to resign.

The CAC members, occasionally shouting and screaming, vented their wrath first on Moniz and Krebs. "[They] did not consider me sufficiently high up in the DOE food chain to waste much time on mauling." Dehmer wrote down some of the recurring themes of the evening:

- This was a purely political decision.
- Terminating the EIS disenfranchised the community.
- DOE can't be trusted.
- You never told us that "programmatic" and "budgetary" concerns would be considered. These are all red herrings to divert us from the fact that this was a purely political decision. (See number 1 comment above.)

- The secretary met with scientists, rich people from the Hamptons, and celebrities, but he never agreed to meet with the rank-and-file members of the BNL community even though there were repeated requests to do so.
- What's going to happen to the workers at the HFBR?
- DOE did a lousy job announcing the decision. (Yep, true.)

Dehmer ended the account by saying, "This was an, um, interesting learning experience for me."[24]

In the next few days, CAC members weighed in to reporters. "Washington took public participation out of our hands," said Mannhaupt. "I think everybody's pretty angry that nobody was informed."[25] A Brookhaven scientist said, "It's a hell of a way to run science policy, and a hell of a way to work with a community."[26] Don Garber, a member of the CAC and past president of the Affiliated Brookhaven Civic Associations, said, "The way Richardson did this [the termination] was obscene."[27] Marburger managed to talk nearly all of the 39 CAC members out of resigning. But one, Richard Raskin of The Pines, an environmental and civic organization, did resign, saying that the "decision-making process has been by-passed." Raskin added, "I believed that decisions would ultimately be made on the basis of science, economics, the needs of the country and the needs and priorities of the community. It is obvious that I was mistaken."[28]

In a memo to lab employees the following Monday, Marburger told them that politics was one element in the HFBR's fate: prominent politicians had decided to oppose the restart and others stood by to let it happen. But, he continued, "DOE's failure to release a draft EIS that could lead the public discourse in the right direction" was another critical element (some people managed to get their hands on copies of the draft after the HFBR termination announcement).

> Like Ed Lowe, most people had only skimmed the news reports. They saw the HFBR as a threat to public health, and their minds would not be changed absent a clear analysis by a credible third party. The EIS was the only thing that came close, and it might have provided the basis for an informed discussion. Would it have been enough to encourage our Congressman's colleagues to deny him the opportunity to prevent restart? I do not know. But last Monday's decision closed the issue.

Marburger continued, "Funding for DOE programs and facilities comes about through a complicated interplay between politics, science and bureaucracy. All three played major roles in the HFBR affair, and science lost. . . . The pain of losing a valuable instrument is very deep, and I feel it keenly, especially as the manner of its loss was so absurd." Marburger ended the memo in a way typical for him, on an optimistic and inspirational note directed at his specific audience. "For me, however, this pain is a stimulus to even greater effort. The will to discovery is indomitable. This Laboratory will prevail."[29]

* * *

In an interview with the *Southampton Press* and other papers, as well as on talk shows, Brinkley took full credit for shutting down the HFBR.[30] STAR members also credited her: "Years later," Brinkley recalled, "I feel a sense of pride in my part in closing the reactor; in helping clean up the air and water of this precious place out here, and I played a tiny part in assuring that other children don't come down with rhabdomyosarcoma or other soft tissue cancers."

On August 18, 2001, Richardson came to Long Island to accept an award from STAR—which specifically cited his role in "permanently shutting down Long Island's High Flux Beam Reactor—handed to him by Brinkley herself at a pop concert given by her ex-husband Billy Joel.

Julius Hastings was more intimately involved with the HFBR than any other scientist. He had joined the lab shortly after it opened in 1947 as the first person with a PhD in its Chemistry Department,[31] hired by department chair Richard Dodson, who helped guide Hastings into a career investigating the magnetic properties of materials. He had helped create the HFBR and was one of the six patent-holders on its design. After it opened Hastings used neutron diffraction to determine magnetic structures of ferrites, magnetic materials that do not conduct electricity, which are both industrially important and scientifically interesting. He was a fixture at the HFBR, generously sharing equipment, techniques, advice, and information with others on the experimental floor and in the Physics Department, and he worked at the reactor continuously even after he retired in 1987. He reacted bitterly to the reactor's closure, appalled both by the decision and by the lack of integrity in

STAR-STUDDED NIGHT WITH BILLY JOEL

For the second year in a row, STAR was honored to present an Evening with Billy Joel, a benefit Master's Class held at Southampton High School. At the August 18th show, Long Island's own Piano Man charmed the audience not only with his incredible musical talents but with his crisp, self-deprecating wit as well.

Before the concert began, STAR chairman Peter Cook and his wife, STAR's environmental activist-extraordinaire Christie Brinkley, awarded former

Former Secretary of Energy Bill Richardson receiving STAR Award from Christie Brinkley

Energy Secretary Bill Richardson the First Annual STAR award. Former Secretary Richardson received this award for permanently shutting down Long Island's High Flux Beam Reactor and passing the Energy Employee Illness Compensation Act Program which finally compensates nuclear workers for radiation illnesses. *In accepting his award Mr. Richardson stated that STAR is making sure that we have an energy-efficient future.*

Figure 7.3

STAR flyer showing former DOE Secretary Bill Richardson receiving an award from Christie Brinkley.

the process by which it had been made. "It would have hurt me less if there had really been something wrong with it," he said.

Ten years after the HFBR's termination, Hastings spoke at a memorial service for Dodson. Hastings said a few words about Dodson, but then unexpectedly shifted focus. "Let me switch gears [and] say a few words of which some may be impolitic. I stand before you with a heavy heart." He proceeded to describe his anger and depression at the role that the spinelessness of administrators, the ambition of politicians, the mendacity of activists, and the profiteering of the media had played in the HFBR's shutdown. He was angry that the advice of scientists who had built the reactor, who understood its lack of danger, and who continually monitored its performance meant nothing when set against the influence of an actor and a model—a supermodel, to be sure, but whose profession, as some of his colleagues angrily put it, was to serve as a visual aid for advertising commercial products. The ability of social agendas without a scientific foundation to terminate a well-functioning scientific instrument whose purpose was to promote and protect human welfare, Hastings feared, meant the entire way of life that he and his colleagues were used to was dying. As a *Newsday* reporter put it, the HFBR episode was a "canary in the coal mine" for the current widespread resistance to scientific findings as is embodied in things like climate change and vaccination denial.

"No serious attempt was made [by the DOE]," Hastings continued, "to quiet the fears of the surrounding community and assure them that the leak presented no health hazard, that it could and would be confined to the laboratory as the latest findings confirm." With great fanfare, the DOE ordered Brookhaven to install a remediation system to recycle the supposedly contaminated water in an attempt to "protect" the community. Since then, the DOE had quietly terminated this remediation system as unnecessary and a waste of money—as a sham. "Since that incident, a pall has descended over Brookhaven," Hastings said. "[It is] a sad legacy."

If this was the fate of the HFBR, Hastings concluded, then social or political agendas could terminate any scientific undertaking no matter how essential for public welfare and safety. This would lead to a different, less enlightened, less safe, human world.

He ended by quoting the poet Dylan Thomas:

Do not go gentle into that good night,
Old age should burn and rave at close of day,
Rage, rage against the dying of the light.

Hastings sat down without a further word.

* * *

The battle over the HFBR had changed character over three years. In 1996, before the leak was detected, the reactor was only the sporadic concern of a small antinuclear group; community concerns focused on potential chemical contamination seeping from other parts of the laboratory's site. In January 1997, the discovery of the tritium leak from the spent fuel pool triggered widespread concerns about the plume's direction, apparent growth, and potential hazard, but it was not found to pose a health threat to employees or the community. In May, the DOE's termination of AUI's contract, and the establishment of an EIS before the reactor's restart, sparked concerns about the possible risk posed by the reactor. In the fall of 1997, political action coupled with a powerful and well-funded antinuclear group, politically and media savvy, made the issue whether the HFBR should exist at all, and targeted the EIS as the Achilles' heel. The battle ended with the fate of the HFBR—safely built and operated, and vital to the U.S. scientific program—entangled in budget issues, the refocused attention of administrators, celebrity influence, and simple exhaustion having nothing to do with the leak at all. "In retrospect," *Physics Today* lamented, "politics and protest prevailed."[32]

Of course it "was a political decision," Bill Smith said, "but it doesn't matter. The important thing is, it's closed."[33]

RETROSPECTIVES

"There but for the grace of God," Fermilab's Judy Jackson had said in comparing BNL's issues to Fermilab's. For this reason, while it is difficult to draw easy lessons from an episode as complex and dramatic, it is important to try. Here are some retrospectives from key players.

LYLE SCHWARTZ

In January 1998, Lyle Schwartz wrote the following "Lessons to be Learned from Brookhaven":

> Whatever you do don't locate your Laboratory and its nuclear reactors over a single-source aquifer on an Island which has in recent years rejected nuclear power after a bitter political fight.
>
> If you must do so, invest very heavily in community relations, building a strong local support group for the Laboratory.
>
> Plan to have no incidents.
>
> But if you must have an incident, try not to have it in areas of ES&H.
>
> And if you do have an incident in ES&H, let it not be in anything nuclear.
>
> And, of course, try not to let any of this happen while you are in the process of changing the President of your management organization.
>
> It should go without saying that you should have done nothing in the past to alienate the political support of your local Congressman and Senator.
>
> Finally, if you have failed to heed the above advice, whatever you do, don't let your mistake occur at a time when a newly appointed Secretary of Energy

needs to make a statement about his management strength and commitment to the environment.

What was the incident at BNL? In retrospect, it is clear that insufficient sensitivity to community relations was a big part of the problem. It also became clear in the self-examination process carried out at BNL that there was a less than satisfactory decision process for prioritizing ES&H activities which resulted in inadequate priority for an action which had already been agreed to with the local community. AUI was actively engaged in an effort to improve community relations, but the discovery of the tritium plume intervened and politics took over. The Management Systems Improvement Plan initiated by AUI and committed to by the new contractor, BSA, will go a long way toward improving both community relations and internal ES&H systems. It is important to emphasize once again as we do in all public communication, that at no time was there a real threat to the health or safety of any BNL employee or member of the local community. Perception of threat is all that mattered, and that perception of threat was real on the part of many of the most vocal members of the local community.[1]

JOHN H. MARBURGER III

In September 1997, during a break at a BSA boot camp, Jack Marburger drafted what he might say to Brookhaven employees at his first meeting with them should BSA win the contract. It began:

> The organization that brought me here today, Brookhaven Science Associates, came into existence as the result of a crisis in the laboratory precipitated by the unexpected termination of the contract between DOE and AUI. What caused this crisis? Was it a random action, driven by politics and taken without regard to the value of the science performed here? I do not think so. I think it was a catastrophe in the modern technical sense—an abrupt adjustment of a system whose parameters had changed almost imperceptibly into a region of instability.
>
> What has changed for Brookhaven are the conditions of the society in which it functions. In an odd sense, the very excellence of BNL's science insulated it from those changes, and made it possible to continue to operate, at least for a while, as if they had not taken place. Society was willing to buy BNL's argument, up to a point, that good science is the bottom line, and that the legalistic

mechanisms of accountability being implemented elsewhere were an expensive luxury whose marginal benefit to society could not balance the reduction in scientific output necessary to create it here.

But the laboratory was operating unwittingly in a region of imbalance with its society. It was only a matter of time before a fluctuation would cause a catastrophic readjustment. . . . If it had not occurred, the lab might have evolved by steady internal change slowly back into balance with its externals. Or it could have persisted in its metastable manner of operation and grown vulnerable to disruption by even smaller perturbations. . . . What we need to do is analyze carefully the conditions under which excellent research may be performed in the actual conditions of the society in which we find ourselves, and create those conditions here at the Brookhaven National Laboratory.[2]

JOHN SHANKLIN

After taking Covello's communication course, Shanklin wrote the following bullet points:

- In the science arena independently verifiable Truth is Reality, in the policy arena Perception is Reality.
- Have a high ranking official (the Director would be ideal) respond to concerns, put a face on faceless government, build a relationship. Say "Hi I'm the director if you have concerns call me personally at 344 xxxx."
- Respond immediately and truthfully to media coverage that is incomplete or inaccurate.
- Seek professional (non-scientist) assistance to sample the concerns of the community and re-poll to determine if the response has addressed the concerns.
- Communication is a learned skill. Sending even the most brilliant scientist to speak on behalf of the lab without being trained how to communicate with non-scientists almost always makes matters worse.
- Some "stakeholders" matter more than others. Build long-term relationships with key players, politicians, celebrities, and community leaders. Involve them in the decision-making process to avoid "Us versus Them" situations.[3]

TARA O'TOOLE

"The BNL tritium leak was not just about that particular lab—we were trying to change the entire DOE safety and environmental culture and demonstrate to the public and Congress that the labs and weapons sites could be responsible stewards."[4]

When asked if the DOE could have handled the episode better, O'Toole hesitated. "No," she finally replied.[5]

MARTHA KREBS

When asked what might have prevented the firing of AUI and the termination of Brookhaven's reactor, Martha Krebs hesitated. "Trusting relationships," she finally said. "Trusting relationships between scientists, politicians, administrators, and community members."[6]

JAN SCHLICHTMANN

A quarter of a century after the event, Jan Schlichtmann remarked, "Maybe they [STAR] got their goal, shutting down the reactor. But it was not a constructive model."[7]

PETER STRUGATZ

"All advocacy didn't seem to be as effective as Christie Brinkley visiting Bill Richardson in DC."[8]

ERNEST MONIZ

"When you are a politician, and you are seeing Congress pushing back, and you are seeing Long Island pushing back, you know it's going to be a hard slog. If you want to reopen it [the HFBR], maybe you'll get there, but it's a hard, hard slog. When it comes to being able to start a major facility, and your own Congressman is arguing against it, you haven't done your job."[9]

SCOTT CULLEN

Twenty-five years later, Cullen said:

> It ended better than it might have. Dr. Marburger was the right person at the right time to help salvage the lab and repair what needed to be repaired. The right person like him can make a huge difference. He went out of his way to try to thread the really difficult needle. Not everybody in a position like that has the temperament or personality or the inclination to do that. From the lab's perspective he should be held as a hero. There was a period of time where everybody was all about, "We've gotta shut this place down, there's no way it should exist in the middle of Long Island." He deserved a good deal of the credit that that didn't happen.[10]

GEORGIA TECH REPORT

In "Contractor Change at the Department of Energy's Multi-program Laboratories," the Georgia Tech's 2001 report concluded:

> Generally speaking, the driving forces behind contractor change are always either political or managerial. While there is obviously some consideration to performance of the laboratory science, technology and national security missions, none of the bid processes has been driven by this consideration. To be sure, increased security, environmental preservation, and cost containment are vital. But if new management increases these and other important ancillary values at the expense of the core mission of the laboratories, contractor change may serve the wrong values. While we have no clear-cut evidence that this has occurred, many of our respondents perceived that contractor change resulted in an elevation of managerial efficiency values over scientific and technical values. This issue requires closer scrutiny than this limited study has been able to provide.[11]

WILLIAM D. MAGWOOD IV

Years later, after becoming Director-General of the Nuclear Energy Agency, Magwood reflected:

> In the nuclear sector, there's a tendency to try to stay out of trouble by flying under the radar, and simply not be noticed. The downside is that, when

something goes a little wrong, you have no capital with the public or Congress to be resilient. The HFBR was a very important and very valuable reactor, and no one outside the community really knew about it. People in New York didn't know about it. They didn't understand what it was doing. So after the leak all they heard were these celebrities saying, "It's hugely dangerous!" People didn't hear the benefits and all the fantastic research into materials and other things. Maybe in hindsight if we were better at talking about these facilities and programs in the normal course of business, instead of talking just within the technical community, we'd have had more resilience with the public—so that when there is a blip, a small thing that goes wrong, the public wouldn't react so strongly. Maybe his [Richardson's] judgment was that this wasn't a battle worth fighting. We didn't like the decision. But it was Richardson's to make.[12]

BNL MEDIA RELATIONS CHRISTMAS PARTY 1999

A new last stanza to "'Twas the Night Before Christmas":

O, HFBR, your fate is sealed
To be decommissioned, with no appeal!
Your name struck from the scroll of science tools,
Only to be remembered, perhaps, as a history lesson in our schools.
O'er a short memory period, you became a local political event,
Relegated as the D'Amato–Forbes victory monument!
But in the long term, in the broader scheme of things,
What do you represent?! A renewal . . . a trend?
Or the beginning of an erosive phenomenon, of happenings
To our national science base that we should amend!?!
Let history be the judge.[13]

AFTERMATH

Though many of the individuals who were driving forces in the events of 1997 left their roles in the year or so afterward, they had a lasting impact on the institutions.

AUI: After its contract to manage Brookhaven National Laboratory was terminated after fifty years, its astronomy portfolio was its sole focus ever since, and grew to include several other radio astronomy telescopes besides the National Radio Astronomy Observatory. They had and continue to have an open and trusting relationship with the National Science Foundation, which funds these facilities.

Alec Baldwin: Baldwin rekindled his movie career, and remained an influential Democratic figure despite his calling for the stoning to death of a Republican congressman, a lawsuit from his daughter, several arrests, and other tabloid-fodder confrontations. He became a popular figure imitating Donald Trump, a notorious science denier himself, on the late-night comedy show *Saturday Night Live*. In 2021, on a movie set, he was given a gun used as a prop that for some unknown reason had a live round in it. He accidentally discharged the gun, killing the cinematographer and wounding the director.

Battelle Memorial Institute: Battelle regained its not-for-profit status and greatly extended its participation in management of DOE laboratories by partnering with others. In addition to Brookhaven and Pacific Northwest National Laboratory (sole contractor since 1965), its portfolio includes Oak Ridge National Laboratory (through UT–Battelle, 2000), Idaho

National Laboratory (through the Battelle Energy Alliance, 2005), Lawrence Livermore National Laboratory (through Lawrence Livermore National Security, 2007), National Renewable Energy Laboratory (in partnership with MRI Global as part of the Alliance for Sustainable Energy, 2008), Los Alamos National Laboratory (through Triad National Security, 2018), and Savannah River (through Battelle Savannah River Alliance, 2021).

Christie Brinkley has received numerous awards for her service to health and the environment. Her causes have included opposition to nuclear weapons and power plants in the United States, and campaigns against the poaching of elephants and rhinos in Africa. She is the author of the best-selling book *Timeless Beauty*, starred in the Broadway musical *Chicago*, and, after eight appearances in the pages of the swimsuit issue of *Sports Illustrated*—and on three covers—in 2017 at age sixty-three appeared again a ninth time.

Brookhaven National Laboratory's environmental cleanup activities, begun years before 1997 at the time it was put on the Superfund list, continued and included major issues such as removal of the BGRR air ducts and the removal of the HFBR stack. Old barracks were demolished and modern buildings constructed. The lab was named "Neighbor of the Year" by a local newspaper, and the International Association for Public Participation named the laboratory "Organization of the Year" for "integrating public participation into its operations." Since 1997, its research has been awarded three more Nobel Prizes, for a total of seven. Years later, Brookhaven's cleanup would be regarded as the best cleanup of a Superfund site on Long Island, and as a model for other such sites.[1] The CAC continued with an evolving membership; some members have dropped out, while others remained; Esposito, for instance, stayed on since the beginning, while STAR (as many predicted the antinuclear group would) has dropped out.

Alfonse D'Amato: After the HFBR shutdown, the Senator affected friendliness to the lab; when D'Amato met Samios, it was back to "How the fuck are ya!" The senator was favored to win reelection in 1998 but was defeated. A prime misstep was when he was reported to have called his opponent, Charles Schumer, a "putzhead," a Yiddish vulgarity that was unacceptably crude for politicians at the time.

Michael Forbes voted to impeach President Bill Clinton in 1998 and won reelection that fall. The following July 17 he abruptly switched parties (becoming a Democrat) and fired his staff without informing them in advance (changing the door locks over the weekend) and also without abandoning his conservative positions. In 2000 Reggie Seltzer, a seventy-one-year-old lawyer whose husband had been a BNL chemist, ran against Forbes in the Democratic primary and defeated him by 35 votes out of 12,000 cast. Her success was partly due to the support of BNL employees and partly to the fact that outraged Republicans funneled a quarter of a million dollars into her campaign. Seltzer went on to lose the general election. In 2013 Forbes was ordained clergy in the Roman Catholic Church in Austin, Texas.

Leon Jaroff continued to tweak STAR, and when he met Baldwin at a Democratic fund-raiser Baldwin was stunned to discover that Jaroff was not a Republican. He died in 2012.

John H. Marburger III remained director of Brookhaven National Laboratory until 2001, when he became Presidential Science Advisor and Director of the Office of Science and Technology Policy under George W. Bush. At the lab farewell party from BNL he was presented with a tritium "exit" sign, of the sort that contained more tritium than the plume. Many scientists did not appreciate his accepting a position in a Republican administration. Marburger became the longest-serving Presidential Science Advisor. He died of non-Hodgkin's lymphoma in 2011.

Robert L. Park continued to be candid and blunt in attacking quacks, pseudoscience, and dishonest activists and politicians. In 2000, while jogging, he was hit by a falling oak tree and nearly killed; some of his targets called it "God's punishment." Park then wrote about the episode in his next book, *Superstition: Belief in the Age of Science*. He stopped writing *What's New* in 2012 after nearly three decades. He died in 2020.

Bill Richardson was indeed a leading contender for Al Gore's choice as vice-presidential candidate. "Remember the HFBR!" was not a popular rallying cry impeding his ambitions; rather, his credibility was dampened by such things as a spy scandal at Los Alamos National Laboratory, in which he hounded an innocent Chinese scientist named Wen Ho Lee. As *What's News* put it, paraphrasing Abraham Lincoln, the case against Lee was "thinner than the homeopathic soup made by boiling the shadow of a pigeon that starved to death." Richardson went on to serve two terms as governor of New Mexico, during the first campaign setting a record for the most number of handshakes in eight hours by a politician.

Stony Brook University was accepted into the Association of American Universities (AAU) in 2001. It is still equal partner to Battelle in BSA.

US Neutron Research, which is an important field especially in materials science, continues to experience a dire shortage of facilities for research and isotope production. The fifty-five-year-old Oak Ridge High Flux Isotope Reactor (HFIR) still operates with four beam tubes and twelve spectrometers, while the lower-power reactor at the National Institute of Standards and Technology (NIST) remains a very good facility—but these instruments are oversubscribed by a factor of three. The Spallation Neutron Source (SNS) at Oak Ridge has provided an expanded capability for some neutron research, but certain areas are better done with reactor sources.

Glossary of Acronyms

AEC   Atomic Energy Commission, a predecessor to the Department of Energy

AGS   Alternating Gradient Synchrotron

ATSDR   Agency for Toxic Substances and Disease Registry, a public health agency focused on the effects of hazardous substances in the environment

AUI   Associated Universities, Inc., the founder and operating contractor of Brookhaven until 1998

BESAC   Basic Energy Sciences Advisory Committee

BGRR   Brookhaven Graphite Research Reactor, a reactor that was shut down in 1968

BMRR   Brookhaven Medical Research Reactor, shut down in 2000

BNL   Brookhaven National Laboratory, one of the DOE national laboratories devoted to research in multiple fields of science

BSA   Brookhaven Science Associates, the contactor chosen to operate BNL beginning in 1998, composed of Stony Brook University and Battelle Memorial Institute

CAC   Community Advisory Committee, a group of community members that provides advice to BNL

CWG   Community Work Group, a group of concerned citizens supported by AUI to oversee BNL environmental issues

DOE   Department of Energy, a department of the US government and owner of seventeen national laboratories

EDF	Environmental Defense Fund
EIS	Environmental Impact Statement
EPA	Environmental Protection Agency
ES&H	Environmental Health and Safety
FOB	Friends of Brookhaven, a self-organized group of BNL employees supporting the laboratory
GAO	Government Accounting (now Accountability) Office, an organization that supplies information and prepares reports for Congress
HFBR	High Flux Beam Reactor, a reactor devoted to research at BNL
IITRI	Illinois Institute of Technology Research Institute
ISME	Integrated Safety Management Evaluation
LIPC	Long Island Progressive Coalition, an activist group devoted to a variety of issues
MSIP	AUI/BNL Management System Improvement Plan
NIH	National Institutes of Health
NIST	National Institute of Standards and Technology
NSLS	National Synchrotron Light Source, an accelerator at BNL
NYSDEC	New York State Department of Environmental Conservation
OER	Office of Environmental Restoration, the BNL organization responsible for Superfund activities
PNNL	Pacific Northwest National Laboratory
RFP	Request for Proposal
RHIC	Relativistic Heavy Ion Collider, a large new BNL accelerator under construction in 1997
RPI	Rensselaer Polytechnic Institute
SCDHS	Suffolk County Department of Health Services
SEB	Source Evaluation Board, the DOE team that evaluates bidders
SNS	Spallation Neutron Source
STAR	Standing for Truth About Radiation, an antinuclear activist group that included high-profile celebrities
VOC	Volatile organic compounds

Acknowledgments

This story was far more complicated than either of us dreamed. Every episode had many more aspects and bizarre dimensions, and was linked to more people and events than we anticipated. Some elements seemed more like plot twists in a work of fiction: politicians consulting celebrities about important scientific issues; a bullying, science-denying actor behaving like nothing so much as the bullying, science-denying politician whom he will become famous for impersonating two decades later; an activist attacking the lab in a hard-core porn magazine; a teenager whose urine was more concentrated than the controversial leak itself; an arch-conservative Republican senator and a former member of "Marxist Feminist Group 1" publicly singing each other's praises; the United States Department of Energy appointing and terminating a management team on the same day; a chaotic competition for a government contract worth hundreds of millions of dollars in which teams drop out hours before deadlines and directors are chosen in days—and then the DOE secretary praising that process as the new way for the agency to do business—and so on. We couldn't make this stuff up. It was not a surprise to us to find that, in 1997, more than one novelist thought of setting Brookhaven as the site of an apocalyptic thriller—and that one such novel was actually published.

If we had known ahead of time how many people we would have to contact; how many documents, emails, and other pieces of paper we would have to go through; and how difficult it would be to keep track of all this material and how much information we would have to assimilate, the task might have seemed hopeless. But the years of the coronavirus pandemic, it turned out, was not such a bad time to contact people. No interviews could be face-to-face, but plenty of people were happy to speak at length by phone.

Crease wound up speaking with politicians, lawyers, scientists, STAR members, LIPC members, EPA workers, DOE officials, Suffolk County health officers, and individuals on the losing and winning bidding teams, among others. We thank all of them for their time, insight, and help. We are grateful that all the principals in this episode we were able to reach were willing and even eager to speak with us, some of them several times. Regrettably, Alec Baldwin's assistant informed us on repeated occasions that the actor could not speak with us, but we were pleased that Mr. Baldwin has left a clear enough record of his thinking in the documentary evidence.

We relied on conversations, interviews, and documents from all these people. Individuals we spoke with and who provided us with information include Joseph Baier, Robert Bari, Miriam Bartos, Frank Bates, Mike Bebon, Marsha Belford, Kevin Billings, Robert Birgeneau, Christie Brinkley, Mike Brooks, fellow Amherst alum Dan Brown, Robert Brown, Helen Caldicott, Bob Casey, Barry Cooperman, Charles Curtis, Frank Crescenzo, Scott Cullen, James Davenport, Sue Davis, Jim Decker, Patricia Dehmer, Ron Delsener, Stephen Dewey, Fred Dombo, Robert Eisenstein, Donald Engelman, Adrienne Esposito, Nanette Essel, Dan Fagan, Paul Falkowski, Deacon Michael Forbes, Joanna Fowler, Jeanne Fox, Kathy Geiger, Peter Genzer, Michael Goldman, Bill Graves, Karl Grossman, Thomas Grumbly, Bill Gunther, Dean Helms, David Heyman, Michael Holland, Elgie Holstein, Judy Jackson, Joshua Johnson, Jean Jordan-Sweet, Ed Kaplan, Michael Kelley, Shirley Kenny, Janos Kirz, Michael Knotek, Lee Koppelman, Martha Krebs, Alan Kuehner, Richard Lahey, Cherri Langenfeld, Terry Lash, T. D. Lee, Mark Linsley, Marianka Louwers, Marge Lynch, Bill Madia, William D. Magwood IV, Joseph Mangano, Frank Marotta, Sean R. McCorkle, Robert McGrath, Robert McNair, Robert Miltenberger, David Moncton, Ernest Moniz, Carson Nealy, Venkatesh Narayanamurti, Ben Ocko, Jennifer O'Connor, Tara O'Toole, Judy Pannullo, Doug Paquette, R. Byron Pipes, Phil Pizzo, Doug Ports, Jordan Rau, Bill Richardson, Mona Rowe, Lisa Rudgers, James Salzman, Nicholas Samios, Jan Schlichtmann,

Howard Schneider, Gary Schroeder, Lyle Schwartz, Stephen Schwartz, John Shanklin, Steven Shapiro, Gerry Shepherd, Steven Silbergleid, Jim Simons, Alice J. Slater, Bill Smith, David Sprintzen, Peter Strugatz, Mary Anne Sullivan, Joseph Topek, Martin Trent, Anne Troutman, Lisa Tyson, Kara Villamil, Paul Vitello, Kate Vodopia, John Wagoner, Diana Weir, Otto White, Leland Willis, Jack Wilson, Jane Yahil, Anne Yoder, Charles Zehren, and Gary Zukas. We apologize if we have left anyone out of this list.

We thank Brookhaven National Laboratory for supplying us with numerous photographs and diagrams. Some individuals gave us access to important documents. Adam Cohen, president and CEO of Associated Universities, Inc., secured us access to the minutes of AUI board meetings. Mona Rowe, Marsha Belford, Kara Villamil, Liz Seubert, and others provided us with piles of material from their files—including printed-out emails, notes, Christmas lyrics and other things that would have been or were being discarded—that only historians would find interesting and important; their work helped keep history alive at the lab. Jack Wilson provided us with portions of his unpublished memoir. Nanette Essel is a historian's dream, an activist who kept and archived every scrap of paper about the activities she was involved with; she scanned and sent us hundreds of documents relating to her community work on environmental cleanup in Suffolk County and her interactions with BNL. The Swarthmore College Peace Collection allowed us access to the Alice Slater papers. Thanks to Alissa Betz and Lisa-Beth Platania, whose excellent and efficient management of the Stony Brook department of philosophy made it possible for Crease to devote time to this book.

Adam Cohen and Mona Rowe read drafts of parts of the manuscript and provided insightful comments. Bill Gunther, Doug Paquette, Kathy Geiger, and others also read and commented on parts of the manuscript. Stephanie Crease, who was working on her own book, tolerated her husband's being on the phone for hours and hours, and for vanishing into his office for days and days. Bond's wife Sandra showed significant patience for times he ignored his surroundings while on his laptop and demonstrated her remarkable filing system by finding a twenty-five-year old negative that became figure 7.2.

We owe much to expert handling by numerous individuals at the MIT Press, including acquisitions editor Katie Helke, acquisitions assistant Laura Keeler, manuscript editors Kathleen Caruso and Stephanie Sakson, designer Emily Gutheinz, and art coordinator Mary Reilly. We owe a great debt to the MIT Press for taking the book as we initially proposed it, with all its outlandish events and extravagant characters, being willing to gamble on a story at once preposterous and prophetic.

Notes

AUI Board Meeting minutes are used with permission. Sources not specified as published items; or as DOE or BNL reports, press releases, or documents; or as issued by AUI or congressional offices, are letters, notes, and other documents in the authors' possession.

PROLOGUE

1. Robert P. Crease, *Making Physics: A Biography of Brookhaven National Laboratory, 1946–1972* (Chicago: University of Chicago Press 1999).

2. P. D. Bond, "The Fiftieth Anniversary of Brookhaven National Laboratory: A Turbulent Time," *Physics in Perspective* 20 (2018): 180–207, https://doi.org/10.1007/s00016-018-0219-x.

CHAPTER 1

1. The drinking water standard applied was federal regulation 40 CFR 141, the Safe Drinking Water Act (SDWA), set by the EPA. That act stipulated radionuclides in water released to a community water system should not cause an annual effective dose that would expose an individual to more than 4 millirem per year, based on a person drinking two liters per day (730 liters per year) of that water. The EPA concluded for tritium that was a concentration of 20,000 picocuries per liter (a picocurie is 0.000000000001 of a curie, or 0.037 disintegrations per second). The DOE's own Technical Standard DOE-STD-1196-2011, referred to in DOE Order 458.1 "Radiation Protection of the Public and the Environment," follows standards adopted internationally (including by the World Health Organization) that set a limit of 80,000 picocuries per liter. Thus the EPA standard is below the DOE's own standards.

2. Doug Paquette to Robert Gaschott, "Re: 96121106.XLS(075–011 & 075–012 tritium, alpha/beta)," January 8, 1997, BNL document.

3. When Brookhaven National Laboratory came into being right after World War II, it was overseen by the Atomic Energy Commission (AEC). The AEC consisted of five commissioners, most of whom were scientists who paid close attention to the national labs. In 1977, the lab's overseer became the US Department of Energy (DOE), a sprawling organization an order

of magnitude larger than the AEC. The DOE had been cobbled together from different governmental agencies, and its relation to the national labs was no longer a close partnership. The DOE managed many more facilities of diverse kinds than the AEC, and to do so it had to impose standardized requirements for performance management and liability. The DOE secretary was a cabinet-level political appointee, frequently not technically trained, and tended to view the national labs as branches of the federal bureaucracy rather than as independent, university-like research organizations. Between DOE headquarters in Washington and the lab were two other DOE offices, an operations office in Chicago and an on-site lab area office. The manager of the area office was Carson Nealy, a career employee at the DOE who had arrived five years previously.

4.   "Unexpected Levels of Tritium Found in Ground Water Monitoring Wells," Brookhaven National Laboratory Reactor Division Occurrence Report CH-BH-BNL-1997-0002, received January 15, 1997, BNL document.

5.   For later studies of the process, see T. Sullivan, M. Hauptmann, and W. Gunther, "Protecting People and the Environment Lessons Learned in Detecting, Monitoring, Modeling and Remediating Radioactive Ground-Water Contamination," Brookhaven National Laboratory Office of Nuclear Regulatory Research, manuscript completed December 2010, date published April 2011, US NRC United States Nuclear Regulatory Commission NUREG/CR-7029.

6.   See note 1.

7.   These investigations and their results are discussed in Idaho National Engineering and Environmental Laboratory, Idaho Falls, Idaho, "High Flux Beam Reactor Tritium Source Identification," vols. 1 and 2, 1997.

8.   One and a half miles is about 7,500 feet; with the groundwater moving about a foot per day, it would take about 7,500 days (20 years) to get to the edge, so with a 12.3-year half-life the total amount of tritium would not only be diluted but would be down by a factor of about four due to its decay.

9.   "Neutrons for the Next Century: The HFBR Upgrade," BNL document.

10. The DOE allowed internal communications, and Samios informed employees on Friday, January 17. N. P. Samios to All Employees, Brookhaven National Laboratory Memorandum, Brookhaven National Laboratory Memorandum, January 17, 1997, BNL document.

11. The Geoprobes were steel pipes an inch and a half in diameter hammered into the ground. When the pipes reached the desired depth at or below the groundwater and cleared of dirt, they were attached to pumps to send water samples to the surface. These samples were sent to an on-site testing center that routinely checked the site's air and water samples, as well as such things as the periodic urine samples required of certain employees, including security guards, to make sure they were drug-free.

12. "Handout, Meeting with Regulators," January 17, 1997, BNL document.

13. These agencies included the US Environmental Protection Agency (EPA), the New York State Department of Environmental Conservation, the New York Department of Health,

and the Suffolk County Department of Health Services; "HFBR Tritium Plume Recovery Plan," January 29, 1997, BNL document.

14. "Brookhaven Lab Notifies Staff of Groundwater Contamination on Site," BNL Press Release no. 97-05, January 18, 1997.

15. Bob Miltenberger to Doug Paquette, January 25, 1997; Doug Paquette to Bob Miltenberger, January 25, 1997, BNL document.

16. "HFBR Tritium Plume Recovery Plan," January 29, 1997, BNL document.

17. Letters from M. S. Davis to Carson Nealy, January–March 1997, BNL document.

18. "Update on Tritium Found in BNL's Groundwater," Brookhaven National Laboratory Memorandum, N.P. Samios to All Employees, January 31, 1997, BNL document.

19. James Salzman, *Drinking Water: A History*, rev. ed. (London: Overlook Duckworth, 2017), 20, 135.

20. Around twenty thousand years ago, retreating glaciers left a ridge of rock, gravel, and sand off what is now Connecticut's shore. By ten thousand years ago, rising seas had turned that ridge into a low-lying island about 120 miles long and 20 miles across, stretching northeast from what today is Manhattan parallel to the Connecticut coast. Native Americans lived on Paumanok, as many called it, even before it was an island. Europeans, arriving in the early seventeenth century, pushed aside the Native Americans, renamed it Long Island, and carved it into counties. They named the easternmost one Suffolk—where Brookhaven is located—after one of the easternmost counties of England.

21. Lab scientist Jan Naidu of the Safety & Environmental Protection Division created a groundwater model to display how an aquifer works, in the form of a huge Plexiglas tray with a bed of wet sand (that's Long Island). When tilted, the water tends to flow in a certain direction (south, to the Atlantic Ocean); drops of colored liquid (the tritium) slowly flowed with the groundwater, while colored pieces of particles (chemical contaminants) tended to stay put. *Brookhaven Bulletin*, August 30, 1996, 2.

22. M. Rogers, *Acorn Days: The Environmental Defense Fund and How It Grew* (New York: Environmental Defense Fund, 1990), 44. Their political activism in other areas also caused friction, in the form of promoting groups such as Planned Parenthood and causes such as black activism. Boyce Rensberger, "Bellport: Friction with the 'Lab People,'" *New York Times*, June 13, 1971, section BQ, 100.

23. In 1980, Congress passed the Comprehensive Environmental Response, Compensation, and Liability Act, known as CERCLA or the Superfund program. Administered by the US Environmental Protection Agency (EPA), it was designed to identify and clean up areas of hazardous waste dumps. Potential Superfund areas were ranked in several categories. One was the "toxicity and waste quantity" of the contaminants, while another had to do with the "people or sensitive environments" potentially affected. Brookhaven scored low in the first category but high in the second, because Long Island's drinking water was primarily supplied by a single aquifer, a "sole source aquifer." The lab was put on the list primarily because of chemical

contamination—pesticides and cleaning agents—from previous disposal practices by the US Army and the laboratory. The lab's score was relatively low—39 out of 100, with a score of 28 sufficient to be put on the Superfund list in 1989. At the end of 1996 Brookhaven was one of 1,227 Superfund sites nationwide, 154 of them federally owned and 20 of them on Long Island. Since the lab is a US Department of Energy (DOE) facility, the DOE was responsible for the Superfund cleanup costs, and its Superfund cleanup is managed by the DOE's Office of Energy Research (the name has been changed more than once, and is currently the Office of Science.)

24. The lab had changed as army barracks were taken down and new facilities constructed, and historical records and the memories of retired scientists were consulted as part of the Historical Legacy survey. Mona Rowe located veterans who had been trained at Camp Upton, and questioned them for what they remembered of the army's disposal practices.

25. See note 1 for this standard.

26. The scope of the Operating Units (OUs) was as follows: OUs 1 and 3 addressed former on-site landfills and other places used to dispose of chemicals, which included chemical plumes dating from Camp Upton days, which were the oldest, deepest, and highest concentration plumes. OU 2 was responsible for contamination from the lab's old waste concentration facility and three other areas of concern. OU 4 addressed an oil spill that had taken place at the central steam facility in the mid-1970s. OU 5 addressed material flowing to the Peconic River, which included the waste from Camp Upton's sewage area at the northeast corner of the site, a simple trough and set of trenches that fed into the Peconic. OU 6 was responsible for cleaning up ethylene dibromide, a common pesticide (and gasoline additive) used on Long Island's farms and also by lab biologists, until it was banned for farm use in 1984. In 1995 a single site-wide groundwater project was established to support all the OUs. Two years later, this project became crucial in allowing the size and character of the tritium plume to be understood so quickly.

27. On the Shoreham controversy, see David P. McCaffrey, *The Politics of Nuclear Power: A History of the Shoreham Nuclear Power Plant* (Dordrecht: Kluwer Academic Publishers, 1991).

28. McCaffrey, *Politics of Nuclear Power*, 53–54. The Shoreham protests picked up momentum following the 1979 accident at the Three Mile Island nuclear power plant in Pennsylvania. Two months later, a demonstration at the Shoreham plant surprised even its organizers, attracting 15,000 protestors and leading to 600 arrests. Shoreham went from being a $600 million project with local support to an unpopular $6 billion white elephant, and it was terminated after the facility was completed and on the verge of producing electricity. The abandoned plant—two teal-painted boxy buildings and a concrete cylindrical tower—still stands behind a fence on the North Shore.

29. Mona Rowe to Richard Rosenthall, letter, November 6, 1995.

30. Elizabeth Collins, "Study by Health Department Ties Cancer Risk to Landfills," *Southampton Press*, August 27, 1998.

31. Jonathan Harr, *A Civil Action* (New York: Random House, 1995).

32. The lab was only one of many contaminated sites on Long Island, and far from the worst. In 1998, Dan Fagin, the author of a Pulitzer Prize–winning book on a severe industrial contamination episode in Toms River, New Jersey, wrote a series of articles in *Newsday* entitled "Underground Danger: The Most Widespread Threat to LI's Water Supply? Thousands of Buried, Leaking Oil and Gasoline Tanks." The two articles—"The Dangers Beneath Us: Buried Gas, Oil Tanks Are Leaking by the Thousands, Threatening LI's Aquifers, as NY Eases Cleanup Rules," and "A No-Win Situation: In Backyards Islandwide, the Nightmare Next Door"— reported that over 40,000 gasoline and oil spills, nearly 10,000 of which did not meet state soil and groundwater standards, had contaminated Long Island's aquifer and drinking water. Many spills came from still-leaking tanks on the property of ordinary houses, gasoline stations, and industrial sites. Rankings of Suffolk County polluters at the time by the EPA, the New York State Department of Health, the New York State Department of Environmental Conservation, and the Environmental Defense Fund all put Brookhaven National Laboratory as less of a concern than many local power plants and industries. Environmental Defense Fund, "The Chemical Scorecard," 995; Chris Giordano, "Port Jeff Power Plant, Toxic Neighbor: State Report Calls Plant Second Worst Polluter in Suffolk" ("outdone only by Northport Power Station") (*Three Village Herald*, February 16, 2000). When in 1998 the EPA ranked Suffolk County's Superfund sites in 1997, the lab came in sixth, behind Circuitron Corp, Lawrence Aviation, Smithtown Groundwater Contamination, Computer Circuits, and Tronic Plating. In 1998, a New York State Department of Health study tied high cancer levels to county and state landfills. Reported in Elizabeth Collins, "Study by Health Department Ties Cancer Risk to Landfills," *Southampton Press*, August 27, 1998, 3. When the state Department of Environmental Conservation ranked toxin emitters, Suffolk County locations such as the Northport Power Station, the Port Jefferson Power Plant, Grumman in Bethpage, and Northville in Calverton were the worst. In 1995, when the Environmental Defense Fund ranked polluters, the lab was thirteenth ("Polluter Locator Scorecard," Environmental Defense Fund, 1995). A 1995 environmental survey ranked the lab tenth among the county's areas with chemical contamination above the limit, with only a fraction the effluents of the first ranked.

33. Karl Grossman, "Insiders Blow Whistle on Brookhaven Dangers," *Covert Action Quarterly*, Fall 1996, 53; Karl Grossman, "Suffolk Closeup," *East Hampton Star*, August 7, 1997.

34. The average natural background radiation in the United States is roughly 360 millirem. Today, someone taking a chest X-ray receives about five millirem, and a thallium stress test 600 millirem. Brookhaven National Laboratory's emissions contribute generally less than 1 millirem to off-site neighbors.

35. Andrew P. Hull, "HTO, HTO Everywhere, but Not a Drop Safe to Drink," *HPS Newsletter* 25 (August 1997): 2–3.

36. "U.S. Department of Energy Offers Public Water Hookups to Residences Just South of Brookhaven Lab," DOE News, January 3, 1996; Joe Haberstroh, "Well-Water Alert," *Newsday*, January 4, 1996, 3.

37. Lauren Terrazzano, "Digging Up Residents' Fears: Suspicion Rather than Gratitude for Free Water Hookups," *Newsday*, March 10, 1996, A35.

38. Kathleen Geiger, phone conversation with R. P. Crease, June 22, 2021.

39. Karl Grossman, "Suffolk Closeup: A Patch of Hell," *Hampton Chronicle-News*, January 26, 1996.

40. Karl Grossman, "Brookhaven Lab: Inquiries, Lawsuits in the Works," *East Hampton Star*, January 25, 1996, 1–10.

41. Strelzin, Balzano & Di Pippo, "Dear Neighbor," February 26, 1997. What happened was this: soon after the January 1996 meeting a $1 billion lawsuit, entitled *Osarczuk v. Associated Universities Inc.*, was filed in February 1996 alleging that pollution from the lab had caused a variety of health issues and had also caused a sharp drop in property values. While only an individual was named as a plaintiff in the suit, it was joined by 300 others from areas just south of the lab. They had hoped of a settlement within five to six years, but it has continued unsettled a quarter-century later.

42. Mona S. Rowe to Department and Division Heads, February 9, 1996, BNL document.

43. In October 1996, the Suffolk County Legislature formed the Brookhaven National Laboratory Environmental Task Force, whose charge was the "evaluation of allegations of contamination and risk associated with the operations of Brookhaven National Laboratory." It had three subcommittees: Epidemiology (co-chaired by Roger Grimson and Dawn Triche), Radiation (chaired by Jane Alcorn), and Non-Radiation (co-chaired by Amy Juchatz and Dawn Triche).

44. New York State Department of Health, Bureau of Environmental Radiation Protection, "Radioactive Contamination in the Peconic River: A Review of the New York State Environmental Radiation Monitoring Program Data," September 9, 1996.

45. Joe Haberstroh, "Crisis Sparks a Solution," *Newsday*, February 25, 1997, A3, A65; DOE News, "U.S. Energy Department Offers Additional Public Water Hookups Near Brookhaven National Laboratory," March 19, 1996.

46. "Fish Unlimited Launches Brookhaven National Laboratory Divestiture Campaign," Fish Unlimited Press Release, September 19, 1996.

47. Kate Vodopia to Bob Casey, March 11, 1996, BNL document.

48. Andrew Lawler, "Grassroots Activist Earns Respect—and Learns Some Hard Lessons, *Science*, February 25, 2000, 1384.

49. "LIPC's Community Activity at The Brookhaven Lab," *Long Island Progressive*, Spring–Summer 1998, 27–28.

50. Leland F. Willis to Judy Pannullo, April 2 and 5, 1996.

51. Nanette Essel, phone interview by R. P. Crease, July 7, 2021.

52. Nanette Essel, phone interview by R. P. Crease, July 7, 2021.

53. Civic Advocates Work Group, Mission Statement, March 1996.

54. Adrienne Esposito, phone interview by R. P. Crease, July 7, 2021.

55. Paul Vitello, "Fast, Furious over Nuke Reactor," *Newsday*, December 2, 1997.

56. August Lockwood, "Environmental Group's Acceptance of Grant Money Stirs Up Controversy," *Times Beacon Record*, May 16, 1996, 1, 6; Nanette Essel, phone interview by R. P. Crease, July 7, 2021.

57. Nanette Essel, phone interview by R. P. Crease, July 7, 2021.

58. Bill Smith to Nanette Essel, March 22, 1996.

59. Sue Davis, "An Open Letter to the Community Working Group," May 10, 1996, flyer.

60. Anne S. Baittinger to Nanette Essel and Jean Mannhaupt, letter, March 28, 1997, BNL document.

61. Helen Caldicott, quoted in the *Sacramento Bee*, April 26, 1988.

62. Helen Caldicott, quoted by Theodore Roszak in *The Oregonian*, June 14, 1992.

63. Beth Greenfield, "Enemies of Brookhaven Lab Cite 'Lies, Secrecy, Silence,'" *Southampton Press*, November 28, 1996, 5.

64. Ray Delgado, "Activists Reveal Naked Truth about Nuclear Catastrophes," *San Francisco Examiner*, October 4, 1999.

65. "Stop the emotion," the anchor Ted Koppel broke in on her. "Let's talk about facts." Helen Caldicott, *A Desperate Passion: An Autobiography* (New York: Norton, 1996), 272.

66. Caldicott, *A Desperate Passion*, 324. In one scorching indictment, environmental activist George Monbiot wrote of her, "Failing to provide sources, refuting data with anecdote, cherry-picking studies, scorning the scientific consensus, invoking a cover-up to explain it: all this is horribly familiar. These are the habits of climate change deniers, against which the green movement has struggled valiantly, calling science to its aid. It is distressing to discover that when the facts don't suit them, members of this movement resort to the follies they have denounced." https://www.allaboutenergy.net/energy/item/2570-the-green-movement-has-misled-the-world-about-the-dangers-of-radiation-helen-caldicott-george-monbiot-australia-uk.

67. Beth Greenfield, "Helen Caldicott Issues a Warning," *Southampton Press*, August 15, 1998.

68. Helga Guthy to Nanette Essel, September 8, 1996.

69. Gary L. Schroeder to General Board of Global Ministries, November 20, 1996.

70. Mona Rowe to BNL Management, August 15, 1996, BNL document.

71. Colin Macilwain, "Brookhaven Feels the Heat over Reactors," *Nature* 384 (November 21, 1996): 205.

72. Christmas song lyrics, handout, December 1996.

73. Elizabeth Kolbert, "The Greening of D'Amato, up from Zero," *New York Times*, February 5, 1998, B1.

74. The League of Conservation Voters rated Forbes at 54 out of 100 percent on environmental issues, and D'Amato at 4 percent. League of Conservation Voters, *1998 National Environmental Scorecard* (for 1997), October 1998, https://scorecard.lcv.org/sites/scorecard.lcv.org/files/LCV_Scorecard_1998.pdf.

75. According to a handout distributed shortly after the January 1996 meeting, "Congressman Mike Forbes does not live anywhere near here. Instead of helping get the millions of gallons of cancer out of our water he has been putting millions of dollars in his pocket." "Water, Water, Cancer, Cancer," handout, February 1996.

76. Joe Haberstroh, "Tritium Levels Rise: New Concerns at Brookhaven Lab," *Newsday*, February 1, 1997, A7.

77. Karl Grossman, "Future of B.N.L. Called 'In Jeopardy,'" *East Hampton Star*, February 27, 1997.

78. The three weapons labs are Los Alamos National Laboratory, Lawrence Livermore National Laboratory, and Sandia National Laboratories.

79. E. Marshall, "Tiger Teams Draw Researchers' Snarls," *Science*, April 19, 1991, 366.

80. Tara O'Toole, phone interview by R. P. Crease, September 14, 2020.

81. Tara O'Toole, phone interview by R. P. Crease, September 14, 2020.

82. The GOCO style relation between the national labs and the government went back to their founding. When the national laboratory system was being worked out in 1946, Congress considered two proposals for managing the labs. One, the May–Johnson bill, retained military control of atomic energy facilities and saw the government as managing the labs similar to the way it managed the Post Office or Tennessee Valley Authority. A rival proposal, the McMahon bill, saw the labs as run by civilians, with independent contractors serving as buffers between the government and the labs. The GOCO system would allow the labs to be federally funded but operated like academic institutions. While the May–Johnson bill was expected to pass easily, scientists vigorously fought it, and succeeded in getting the McMahon bill passed. The DOE's effort to weaken the GOCO system increased after the Watkins "Tiger Teams."

83. Bernard Manowitz to Warren Winsche, April 19, 1948, BNL document C-1698.

84. Brookhaven scientists also studied greenhouse gases and pollutants in the food chain. In a "gamma field," lab biologists investigated radiation on plants—and after the 1986 Chernobyl nuclear power plant explosion in the Soviet Union, the lab's data helped scientists determine from aerial photos the radiation doses involved.

85. BNL Progress Report, 7/1–12/31, 1948, BNL document.

86. The DOE rated AUI's performance with five of 15 measures "outstanding," six "excellent," and three "good," and its performance overall was "excellent." H. C. Grahn [reminding the DOE official of the ranking] to C. Nealy, December 27, 1996, BNL document.

87. AUI Board Meeting minutes, January 29–30, 1997.

88. Barry Cooperman, phone interview by R. P. Crease, October 4, 2019.

CHAPTER 2

1. "Evaluation of Alternatives for Tritium Remediation," memorandum, Bill Gunther to Distribution, February 9, 1997, BNL document.

2. Sue Davis, interview by R. P. Crease, Brookhaven National Laboratory, June 1998.

3. Bill Gunther to Doug Paquette, "Tritium Pumpage, Geoprobe," February 14, 1997, BNL document.

4. Bill Gunther, phone interview by R. P. Crease, April 24, 2020. The final report: Idaho National Engineering and Environmental Laboratory, Idaho Falls, Idaho, "High Flux Beam Reactor Tritium Source Identification," vol. 1, May 16, 1997; vol. 2, July 1997, BNL document.

5. "Status of Investigation of the Spent Fuel Canal Using Horizontal Wells," April 16, 1997, in "HFBR Tritium Source Identification," vol. 2.

6. Henry Grahn to R. P. Crease, "Cleanup Money," email, July 1, 1998, BNL document.

7. Bob Miltenberger to Kara Villamil, email, February 15, 1997, BNL document.

8. Andrew Lawler, "Meltdown on Long Island," *Science*, February 25, 2000, 1386.

9. Karl Grossman, "Suffolk Closeup: Stakes Get Higher," *Southampton Press*, March 6, 1997, 8.

10. Lisa M. Hamm, "Groundwater around Brookhaven Reactor Tainted by Radioactive Tritium," Associated Press story draft, February 1, 1997.

11. John Rather, "High A-Levels in Water Spur Brookhaven," *New York Times*, January 26, 1997, Long Island section, 4.

12. Pete Maniscalco, "Brookhaven Lab Nuclear Reactors Crucify Mother Earth: Lenten Prayer Vigil at Brookhaven Lab," February 11, 1997.

13. Media Relations, BNL, "Media Activity Summary, HFBR Tritium," February 18–28, 1997, BNL document.

14. Lisa M. Hamm, "Long Islanders Say Research Lab Is Polluting Water Supply," *Grand Rapids Press*, August 3, 1997, A12.

15. *Suffolk Life*, February 5, 1997. The article on the inside, though, spoke only of the shutdown of the HFBR.

16. Mona Rowe to Phyllis Ballard, September 24, 1997, BNL document.

17. Bill Smith, phone interview by R. P. Crease, July 15, 2021.

18. Karl Grossman, "Banners Blast Lab," *East Hampton Star*, May 29, 1997, 1–11.

19. Stephen Dewey, phone interview by R. P. Crease, March 26, 2021.

20. Stephen Dewey to Mona Rowe, February 11, 1997, BNL document.

21. Mark Crawford, "Tritium Leak Threatens Neutron Source's Future," *New Technology Week* 11, no. 7 (February 18, 1997): 1, 7, 11.

22. Steven Baker to Doug Paquette, "Groundwater Material for Tara O'Toole," February 17, 1997, BNL document.

23. "Points for Conversation with T. Lash on 2/12/97, Announcement of Extension of Public Water," BNL document.

24. Alfonse D'Amato and Michael Forbes, "Media Advisory: D'Amato, Forbes to Make Announcement Monday Regarding Brookhaven Lab and Local Community," February 23, 1997.

25. Quoted in the *Brookhaven Bulletin*, February 28, 1997.

26. Tara O'Toole, phone interview by R. P. Crease, September 14, 2020.

27. "Suffolk County Legislature, Public Hearing, Brookhaven National Laboratory, February 20, 1997," transcript.

28. Forbes and more local politicians, including legislator Michael Caracciolo, New York State Attorney General Dennis Vacco, and others, sounded similar notes.

29. "Suffolk County Legislature, Public Hearing, Brookhaven National Laboratory, February 20, 1997," transcript. Karl Grossman's description of the meeting: "Suffolk Closeup: Stakes Get Higher," *Southampton Press*, March 6, 1997, 8.

30. For more on the US neutron facilities, see John J. Rush, "US Neutron Facility Development in the Last Half-Century: A Cautionary Tale," *Physics in Perspective* 17 (2015): 135–155.

31. The innovative design had to do with the arrangement of its 28 fuel elements. The fuel elements of previous reactors, including the BGRR, were separated by moderators—graphite, in the BGRR—which slowed down neutrons enough to continue the fission process. In the HFBR, the fuel elements were placed close together inside a small, bulb-shaped vessel whose moderator was heavy water—hydrogen nuclei containing a proton and neutron rather than just a proton. The HFBR's 28 fuel elements were about two feet long and three inches on a side. Eight times a year, operators shut the reactor for several days and used a crane to pull seven elements out of the core and lower in new ones.

32. Robert P. Crease, "Anxious History: The High Flux Beam Reactor and Brookhaven National Laboratory," *Historical Studies in the Physical and Biological Sciences* 32, no. 1 (2001): 41–56.

33. For an overview of research at the HFBR as of 1997, see "HFBR Scientific Highlights," BNL document. The facility attracted theorists as well as experimenters, and had a popular summer program.

34. William D. Magwood IV, Zoom interview by R. P. Crease, September 1, 2021.

35. In 1994, a faulty electrical feed caused a fire during an experiment that resulted in seven individuals being exposed to a small amount of radioactivity. There was virtually no release of radioactive material outside the HFBR. The episode led to some revisions in technical and management practices at the HFBR. US Department of Energy, "Type B Investigation of the March 31, 1994 Fire and Contamination at the Tristan Experiment, High Flux Beam Reactor, Brookhaven National Laboratory, Upton, NY," April 29, 1994, BNL document.

36. A new advanced reactor, the Advanced Neutron Source, was planned for Oak Ridge National Laboratory, but in 1994 Congress terminated it, eventually approving a plan to build a next-generation pulsed Spallation Neutron Source—an accelerator-based neutron source—at that lab.

37. The upgrade: "Conceptual Design Report: HFBR Neutron Beam Lines and Facilities Upgrade," February 1994, Project No. 96-CH-120. The quote is from Stringer, letter to

Krebs, November 22, 1997. See also J. D. Axe, W. J. Brynda, L. Passell, and D. C. Rorer, "Renewing the High Flux Beam Reactor," *Neutron News* 7, no. 1 (1996): 24–26. The Birgeneau BESAC report captured the urgency felt by the US neutron scattering community for a new neutron source but not the anger at how the DOE was responding to their needs. Two days after Birgeneau's letter, Samuel Werner, the president of the Neutron Scattering Society of America, went around the DOE directly to Congress and wrote a blistering letter to the chairman of the US House of Representatives to complain about the DOE's handling of facilities for neutron research. Despite the fact that neutron sources make fundamental contributions in biology, chemistry, and materials science and to understanding the materials inside things like motors, generators, batteries, and recorders, the DOE has allowed US neutron sources to "deteriorate," constructed no new ones, given "no real support" to US neutron scatterers, and is about to surrender the US lead in neutron scattering research. The DOE's failure to support US neutron sources more vigorously threatens to "extinguish a field to which so much of our technical knowledge and economic power owes its very existence." Samuel A. Werner to Robert S. Walker, March 12, 1996. Werner pleaded for action to build a new neutron source, or at least an upgrade—and the HFBR was the only suitable neutron instrument to upgrade.

38. Charles F. Majkrzak to Martha Krebs, January 3, 1996, BNL document.

39. Another source of tritium in the spent fuel pool was condensation on pipes and other equipment, which at first was sent through drains into the sanitary sewer system and to the Waste Treatment Facility. In the mid-1990s, lab officials reduced the amount of tritiated water going out of the building that way and it started to be sent into the spent fuel pool, causing the amount of tritium into the pool to rise—and reducing the amount of makeup water that had to be added.

40. J. M. Hendrie, "Final Safety Analysis Report on the Brookhaven High Flux Beam Research Reactor," BNL 7661, April 1964, no. 7.8.5, "Spent Fuel Storage."

41. "Agreement between Brookhaven National Laboratory and County of Suffolk," signed by Michael A. Lo Grande, County of Suffolk, Nicholas P. Samios, Brookhaven National Laboratory, and a DOE representative, US Department of Energy, September 23, 1987. The agreement read in part, "Recognizing that the environmental concerns of the people of Suffolk County have broadened dramatically in scope and complexity in the past few years, and that there is now a great interest in assuring that the same standards of environmental protection and control are applied to facilities of the Brookhaven National Laboratory as are applied to the rest of the community, and . . . recognizing that the Laboratory shares with Suffolk County a strong and dedicated concern for environmental protection . . . it is the desire of both Brookhaven National Laboratory and the Suffolk County Department of Health Services to move in the spirit of comity to establish the highest practical level of environmental protection for the citizens and lands of Suffolk County from any problems arising at the Brookhaven National Laboratory property, facilities and operations, by making every effort to conform to local, State and Federal regulations related to public health and environmental protection. . . . Therefore, to assist in achieving this high level of protection, it is agreed that . . . the Laboratory will conform

with the applicable environmental requirements of Articles 6, 7, 10, and 12 of the Suffolk County Sanitary Code. Brookhaven National Laboratory and the Suffolk County Department of Health Services agree to work in good faith to achieve, as expeditiously as possible, a reasonable schedule for upgrading facilities as required." The agreement grew out of a "Task Force" appointed by Gregory J. Blass in 1985, which issued a report the following year.

42. Joseph Baier, phone interview by R. P. Crease, October 16, 2020.

43. See chapter 1, note 3.

44. After she arrived at the DOE, O'Toole's first major achievement was a "Spent Fuel Working Group" to visit DOE laboratories and assess the safety of their spent fuel storage facilities. The group was primarily concerned with defense labs such as Hanford, which had by far the most serious problems, but it also visited non-defense labs. During its 1993 visit to Brookhaven, O'Toole's team assessed the HFBR's spent fuel pool. Its final report, signed by O'Toole, stated that "no leakage problems have ever been detected" in the HFBR's canal and that its "one potential vulnerability" had to do with earthquake damage—an extremely remote prospect in New York and especially Long Island. O'Toole's report stated, "The fuel canal is unlined and there is no continuous and accurate way of measuring leak detection. However, alarms for high and low water level are in the control room and the water level is regularly monitored. Records are maintained for canal water additions and thus any increased amounts of canal makeup water can be detected. The canal has been sealed against evaporation about every five years to measure leakage, and no leakage problems have ever been detected. Also, there are ground water monitoring wells near the High Flux Beam Reactor that are sampled twice per year, and no significant amounts of radionuclides have ever been detected." US Department of Energy, "Spent Fuel Working Group Report on Inventory and Storage of the Department's Spent Nuclear Fuel and Other Reactor Irradiated Nuclear Materials and Their Environmental, Safety and Health Vulnerabilities," vol. 2, November 1993, https://www.energy.gov/sites/prod/files/2014/04/f15/Spent_Fuel_Working_Group_Report_1993.pdf.

45. R. Miltenberger to D. Ports, "NRC Information Notice 92-11: Soil and Water Contamination," July 7, 1993, BNL document.

46. Doug Paquette to Mona Rowe, January 25, 1997, BNL document.

47. R. Karol to Files, "Further Information Regarding Estimated HFBR Canal Leak Rate Due to Concrete Permeability," February 25, 1997, BNL document; "Evaluation of the Permeability of Water through the HFBR Spent Fuel Canal Floor Slab," December 4, 1997, BNL document.

48. Doug Ports, phone interview by R. P. Crease, September 17, 2020.

49. "Tritium Source Identification—Canal, Summary," BNL document.

50. Doug Ports, "HFBR Fuel Canal Leak Test Report," March 3, 1997, in HFBR Tritium Source Identification, vol. 2, BNL document.

51. D. Ports, "HFBR Fuel Canal Leak Rate Test Report," Leak Test no. 2, March 20, 1997, in HFBR Tritium Source Identification, vol. 2, BNL document.

52. Charlie Zehren, "At Lab, Bucket of Trouble," *Newsday*, November 14, 1997, A3, A70.

53. A complete set of data from all the wells characterizing the plume can be found in M. H. Brooks to Carson L. Nealy, "Transmittal of Data for the Tritium Remediation Project," June 1, 1997, BNL document. A review of the models can be found in Bill Gunther, "Assessment of the HFBR Tritium Plume Models," Spring 2005, BNL document.

54. The 68,000-gallon spent fuel pool's tritium concentration varied over time, averaging about 130 microcuries per liter. That indicates a total of about 34 curies in the pool. The leak rate was measured between six and nine gallons per day. Assuming that rate was constant for twelve years, a total of between 26,000 and 39,000 gallons leaked, or approximately one-half the total volume of the spent fuel pool. Using the actual leak rate implies that the plume contained between 13 and 20 curies. But during those twelve years the tritium was decaying, so the net tritium in the plume was reduced to between 9.4 and 14.4 curies. To be conservative, we have chosen a value of about 15 curies for the total amount of tritium in the plume.

55. Bernard Manowitz to Warren Winsche, April 19, 1948, BNL document C-1698.

56. Gary Schroeder to Doug Paquette, handwritten notes on Steve Baker, "Tritium Plume Model Evaluation," April 20, 1997.

57. *Newsday*, April 9, 1997, B26.

58. Patricia Dehmer to R. P. Crease, email, November 12, 2021.

59. Joe Haberstroh, "Tritium Levels Rise," *Newsday*, February 1, 1997; "Lab's Toxic Plume Wider than Thought," *Newsday*, February 7, 1997.

60. Bond notes of Lash-O'Toole visit to BNL, March 4, 1997.

61. "March 4 Update on Groundwater Contamination at Brookhaven National Laboratory," faxed press release, March 4, 1997, BNL document.

62. John Shanklin, "Questions on BNL," *The Sound Observer*, April 11, 1997.

63. Information Service Division, Brookhaven National Laboratory, "Why Do We Continue to Have 'Revelations' about Contamination and Other Problems at BNL?," March 20, 1997, BNL document

64. Miltenberger produced a criterion for what measurements would count as data, but it did little to stem criticism. For instance, in R. Miltenberger to M. S. Davis, February 12, 1997, BNL document; M. S. Davis to Mary E. Hibberd, "Background on BNL Analytical Lab, 3/21/97," March 20, 1997, BNL document.

65. E.g., N. P. Samios to All Employees, "Historical Data on Drinking Water Well South of HFBR," March 10, 1997, BNL document.

66. Cover editorial, *Newsday*, Currents & Books section, April 6, 1997.

67. Robert L. Park, "Brookhaven: Senator D'Amato Calls for Senate Hearings," *What's New*, March 21, 1997, http://bobpark.physics.umd.edu/WN97/wn032197.html.

68. "Tritium Labeling of Organic Compounds Deposited on Porous Structures," US Patent No. 4,162,142.

69. Liz Seubert, "Having Identified Leading Edge of On-Site Tritium Plume, Lab Plans Fuel Shipments, Groundwater Remediation," *Brookhaven Bulletin*, March 28, 1997, 1.

70. Robert Bari, interview by R. P. Crease, Stony Brook, October 29, 2020.

71. Federico Peña to F. James Sensenbrenner Jr., June 12, 1997.

72. *Brookhaven Bulletin*, March 28, 1997, 1.

73. On March 22, the *New York Times* ran a relatively balanced cover story about Brookhaven's troubles, reviewing events of the past several years, including the chemical plumes, and mentioned that both the lab and independent agencies agree that the tritium posed no health risk. It quoted Robert Casey, head of the lab's safety and environmental division, as saying that the leak "makes it hard for people to accept our other statements that we are concerned and doing a good job." The article also said that, with all the repercussions of the leak, including attacks from D'Amato and Forbes, "Brookhaven has so many problems that a radioactive plume can seem like just another item on a maintenance crew's checklist." Dan Barry and Andrew C. Revkin, "At 50, Brookhaven Lab Is Beset by Problems," *New York Times*, March 22, 1997, 1; also Joe Haberstroh and Liam Plevin, "Down to Business," *Newsday*, April 21, 1997.

74. Joe Haberstroh and Liam Pleven, "Brookhaven Lab Seeks More PR in Top Post," *Newsday*, March 8, 1997.

75. N. P. Samios, "To All Employees of Brookhaven National Laboratory," March 7, 1997; "Brookhaven National Laboratory Announces Retirement of Director," BNL Press Release no. 97-22, March 7, 1997. Hughes's announcement: Robert E. Hughes, "To All Employees of Brookhaven National Laboratory," March 7, 1997, BNL document.

76. N. P. Samios to All Employees, "Review of ESH Decision Making," March 3, 1997, BNL document.

77. Marge Lynch to Lyle Schwartz, Mike Bebon, and Peter Bond, "Bullet Points for Meeting with John Wagoner," May 20, 1997, BNL document.

78. Earl Lane, "Lab Chief's Dueling Legacies," *Newsday*, March 9, 1997, A6, A36.

79. Paul Martin, "To All Employees of Brookhaven National Laboratory," March 20, 1997, BNL document.

80. Peter Bond, "Possible Interim Director," Notes of Meeting, April 1, 1997.

81. Sue Davis, "Brookhaven National Laboratory Tritium Modeling & Conceptual Design Review," April 9–10, 1997, BNL document.

82. Bill Gunther, phone interview by R. P. Crease, April 24, 2020.

83. Tara O'Toole to Peter Bond, May 15, 1998.

84. Associated Universities, Inc., Board of Trustees Meeting minutes, April 16–17.

85. Martha Krebs, phone interview by R. P. Crease, October 20, 2020.

86. Associated Universities, Inc., Board of Trustees Meeting minutes, April 16–17.

87. Peter Bond, "Martha Krebs," Notes of Meeting, April 16, 1997.

88. Peter Bond, Notes of Meeting, April 22, 1997.

89. Schwartz had also arranged for a "PNNL Management Improvement Team" to come to the lab for three and a half days starting on May 6. The team members discussed a 1993 episode at Savannah River in which 7,000 curies of tritium were dumped into the Savannah River, and major spills at Los Alamos much more serious than at those at Brookhaven. But in comparatively sparsely populated New Mexico, the two national security labs—Los Alamos and Sandia National Laboratory—were considered crown jewels and strongly supported by Senator Pete Domenici, while in densely populated New York the national laboratory was a small part of the state's economy and drew less attention from the state's senators—except negatively, when pro-environmental credentials could be had. Schwartz found that the consultants had some relevant lessons but to solve Brookhaven's problems they "must have people who know existing culture—so internal people needed." Peter Bond, "Notes of Meeting with Laity and Rice," March 6, 1997.

90. Carmen MacDougall to Greg Cook, Frank Crescenzo, and Carson Nealy, fax, April 23, 1997. The lab drew up a series of measures to respond to the issues raised by the two reports.

91. Tara O'Toole, phone interview by R. P. Crease, September 14, 2020.

92. Tara O'Toole, Al Alm, Terry Lash, Martha Krebs, and Mary Anne Sullivan, "Memorandum for the Secretary: Response to Management Problems at Brookhaven National Laboratory (BNL)," April 16, 1997.

93. "It would be difficult legally to show that AUI did not make its 'best efforts' to carry out the contract." Tara O'Toole, Al Alm, Terry Lash, Martha Krebs, and Mary Anne Sullivan, "Memorandum for the Secretary," undated.

94. O'Toole, Alm, Lash, Krebs, and Sullivan, "Memorandum for the Secretary."

95. Elgie Holstein, phone interview by R. P. Crease, October 26, 2020.

96. "Associated Universities Board Names Interim Director of Brookhaven National Laboratory," no. 9-43, April 28, 1997, BNL document.

97. Lyle Schwartz, "Press Conference Remarks," April 28, 1997.

98. Tara O'Toole, phone interview by R. P. Crease, September 14, 2020.

99. Joe Haberstroh and Liam Plevin, "Lab Appoints Interim Boss," *Newsday*, April 29, 1997.

100. US Department of Energy, Office of Environment, Safety and Health, "Integrated Safety Management Evaluation of Brookhaven National Laboratory," April 1997, EH2MGT/04-97/02SH.

101. R. A. Bari, D. Gordon, D. Moran, N. Volkow, and C. Meinhold, "Report of the Ad Hoc Committee on Environmental, Safety and Health Decision Making at Brookhaven National Laboratory," April 29, 1997, BNL document.

102. Lyle Schwartz, phone interview by R. P. Crease, September 4, 2020.

103. "Associated Universities, Inc. and Brookhaven Lab Announce Aggressive Response to Dept. of Energy Review," BNL Press Release no. 97-45, May 1, 1997.

104. Peter Bond, notes after phone call with Schwartz, April 30, 1997.

105. In Tara O'Toole's memo summarizing the ISME report for the ES&H office, she reports that it says that only BNL needs to make improvements: "The overall conclusion of this evaluation is that environment, safety and health (ES&H) management at BNL is in need of improvement and significant management attention." Tara O'Toole, "Brookhaven National Laboratory Integrated Safety Management Evaluation Report," May 1, 1997, BNL document.

CHAPTER 3

1. The next day, May 2, the DOE's Brookhaven Group Manager Carson Nealy sent AUI head Lyle Schwartz official announcement of the termination. Nealy to Schwartz, May 2, 1997, BNL document.

2. Frank Crescenzo, phone interview by R. P. Crease, June 5, 2020.

3. Robert Bari, phone interview by R. P. Crease, October 29, 2020.

4. DOE Press Release R-97-032.

5. Each was an experienced manager from another DOE lab. Wagoner managed the DOE's Richland Operations Office, one of his missions being to clean up radiation leaking from the Hanford site's previous defense production work, and in 1994 had worked with the community after a major radiation spill in 1994, which leaked into waters leading into the Columbia River, a leak that was still continuing. Helms had been employed by the DOE and its predecessor agencies for three decades and was the DOE site office manager at the Thomas Jefferson National Accelerator Facility, in Newport News, Virginia. He had been the Tiger Team leader for the 1990 investigation of the Savannah River site.

6. US Department of Energy, "DOE Action Plan for Improved Management of Brookhaven National Laboratory," draft, June 1997.

7. The lab's ES&H performance had already been reviewed twice in April, once by the DOE's ISME team and once by the Bari committee. Each had taken time away from lab employees who were interviewed and asked to compile information and produce documents. In May came three additional reviews, by the EPA, by the state Attorney General's office, and by a Suffolk County Task Force. New York State Attorney General Dennis Vacco sent a review team, expressing annoyance that he had not been informed earlier about the environmental issues at Brookhaven, despite the fact that his office had been kept up to date by the members of his staff on the ISME committee. The Suffolk County Task Force was also examining health and environmental issues possibly caused by the lab.

8. Tara O'Toole, "Brookhaven National Laboratory Integrated Safety Management Evaluation Report," May 1, 1997, BNL document.

9. When Bond later called O'Toole out on the remark, saying it reflected an unfairly antagonistic attitude toward the lab, she lashed out. "[I]t absolutely blows my mind that you, a guy who's

used to integrating lots of data and making judgments about what it means, would latch onto this shred of language, spoken in a rather high-pressure situation and think it somehow is a meaningful signal of intention, intelligence and purpose and somehow a litmus test of the 'goodness' of my relations with BNL." Tara O'Toole to Peter Bond, June 27, 1997. Yet O'Toole's remark about wells saving millions was not improvised in haste but was in her prepared remarks and the DOE press release; "For example, in 1995 BNL management decided to postpone installation of groundwater monitoring wells at the HFBR in favor of 'higher priority' activities, a delay that will ultimately cost tens of millions of dollars." "Secretary Peña Terminates Brookhaven Contract," *DOE News*, May 1, 1997, R-97-032.

10. Hubert Herring, "Diary," *New York Times*, May 4, 1997, section 3, 2.

11. Clifford Krauss, "New Brookhaven Lab Manager Says He's Sure Drinking Water Is Safe," *New York Times*, May 3, 1997, Metro section, 26.

12. Lyle Schwartz, "Message to Employees," remarks on May 1, 1997, BNL document.

13. The EIS was a feature that had been incorporated into the planning process for projects on federal land by the National Environmental Policy Act (NEPA) of 1969. The process worked as follows: first, the government lays out the possible environmental impacts and asks for and reviews public comments either in writing or at public meetings. After that, the DOE draws up an implementation plan that summarizes the results of the scoping, which is also made available for public comment. A draft EIS is then prepared, which also receives public comment, whereupon the draft is revised and a final EIS issued.

14. *Science and Government Report*, May 15, 1997, 1.

15. J. Madeleine Nash, "Scientists Are Discovering the Chemical Secret to How We Get Addicted . . . and How We Might Get Cured," *Time*, May 5, 1997, 69–76.

16. N. D. Volkow, G.-J. Wang, J. S. Fowler, J. Logan, S. J. Gatley, R. Hitzemann, et al., "Decreased Atriatal Dopaminergic Responsiveness in Detoxified Cocaine-Dependent Subjects," *Nature* 386, no. 6627 (1997): 830–833, https://doi.org/10.1038/386830a0; N. D. Volkow, G.-J. Wang, M. W. Fischman, R. W. Foltin, J. S. Fowler, N. N. Abumrad, et al., "Relationship between Subjective Effects of Cocaine and Dopamine Transporter Occupancy," *Nature* 386, no. 6627 (1997): 827–830, https://doi.org/10.1038/386827a0.

17. Hong Li et al., "Crystal Structure of Lyme Disease Antigen OspA Complexed with an Fab," *Proceedings of the National Academy of Sciences* 94, no. 8 (April 15, 1997): 3584–3589, https://doi.org/10.1073/pnas.94.8.3584.

18. AUI Board Meeting minutes, June 11–12.

19. Associated Universities Executive Board Meeting minutes, May 15, 1997.

20. "Statement by Jeanne M. Fox, EPA Regional Administrator," United States Environmental Protection Agency News, May 15, 1997.

21. M. S. Davis to L. H. Schwartz, "ISME Statements on the HFBR Spent Fuel Pool and DOE's Spent Fuel Vulnerability Study," May 12, 1997.

22. O. White, "A Comparison between BNL ISME Report and LANL IOE Report," May 7, 1997.

23. Karl Grossman, "Suffolk Closeup: The Lost Reactor," *Southampton Press*, May 29, 1997, 12.

24. James Sensenbrenner Jr. and George E. Brown Jr. to Charles A. Bowsher, June 3, 1997.

25. They called on the GAO to investigate "1) how the Brookhaven situation developed, including the breakdown of public trust; 2) who was at fault; 3) what conditions and management processes, or lack thereof, led to the tritium incident and to the decision to terminate AUI; and 4) the Department's failure to put in place a management system that has clearly defined authority, roles, responsibilities and accountability for all those involved, from the Secretary down to the operational people at BNL."

26. Robert L. Park, "Brookhaven? General Accounting Office Asked to Fix the Blame," *What's New*, June 6, 1997, http://bobpark.physics.umd.edu/WN97/wn060697.html.

27. Colin MacIlwain, "Brookhaven Contractor Is Sacked over Tritium Leak," *Nature*, May 8, 1997, p. 114.

28. R. E. Gerton, "Accident Investigation Board Report on the May 14, 1997, Chemical Explosion at the Plutonium Reclamation Facility, Hanford Site, Richland, Washington—Final Report," US Department of Energy, July 26, 1997, DOE/RL-97–59, https://doi.org/10.2172/325379.

29. Carson L. Nealy to Lyle Schwartz, "Assessment of Performance of Associated Universities, Inc., and Brookhaven National Laboratory for the Period October 1, 1995 through September 30, 1996," May 16, 1997, BNL document.

30. Paul Martin, "Statement to BNL Staff," May 13, 1997, BNL document.

31. Associated Universities, Inc., Executive Committee minutes, May 9, 1997.

32. Associated Universities, Inc., Executive Committee minutes, May 9, 1997.

33. Richard Hames to Linda Rhode, FOIA Officer, May 27, 1997.

34. Lyle Schwartz to Carson L. Nealy, May 20, 1997, BNL document.

35. Carson L. Nealy to Lyle Schwartz, "Termination for Convenience of Prime Contract No. DE-AC-76CH00016," May 29, 1997, BNL document.

36. On June 9, Schwartz wrote a blistering letter denying Nealy's statement that senior DOE officials had spoken frequently with AUI in the months before the decision. Schwartz protested that "there were *no* discussions with AUI management that the Department was considering immediate termination and recompetition of the contract." Schwartz added that the DOE "approval of the interim management team three days prior to its precipitous termination led me to conclude that our corrective actions were appropriate and effective." Schwartz also cited the finding of the DOE Safety Management Evaluation of BNL that "the current Department of Energy and Brookhaven National Laboratory actions to eliminate the source and remediate the tritium contamination have been aggressive and appropriate." Schwartz concluded by requesting the DOE to rescind the termination, calling the decision an act of "bad faith, abuse of discretion and renders the decision to terminate a substantial breach of

contract." Lyle Schwartz to Carson L. Nealy, "Re Termination of Contract No. DE-AC02-76CH00016," June 9, 1997, BNL document.

37. It announced that "Associated Universities Inc. finds it necessary to provide this formal notification to all scientific staff members that their existing scientific appointments will terminate" at the end of AUI's contract with the DOE. Lyle H. Schwartz to Scientific Staff, "Notice of Termination of AUI Scientific Staff Status," May 22, 1997, BNL document.

38. Robert L. Park, "Brookhaven: DOE Announcement Calms Scientists—Sort Of," *What's New*, May 30, 1997, http://bobpark.physics.umd.edu/WN97/wn053097.html.

39. Liam Pleven, "Scientists Setting Sail? Brookhaven's Best Tempted to Leave," *Newsday*, June 9, 1997, A8, A35.

40. Lyle Schwartz, interview by R. P. Crease, June 1998.

41. Peter Yamin to Federico Peña, May 30, 1997, BNL document.

42. "Red Tape Must Not Strangle Good Science," *Nature* 387 (May 8, 1997): 107.

43. Noel Corngold, "BNL Shakeup Tied to 'Usual Medieval Suspects,'" *Physics Today*, November 1997.

44. John Shanklin to R. P. Crease, email, December 29, 2020.

45. Joanna Fowler, phone interview by R. P. Crease, July 21, 2020.

46. John Shanklin, phone interview by R. P. Crease, January 9, 2021.

47. S. E. Schwartz, "Notes from Meeting with Victor Yannacone 5-24-97."

48. Joanna Fowler, phone interview by R. P. Crease, July 21, 2020.

49. Friends of Brookhaven, "Petition," and FOB to John Wagoner, "Recommendations for the Request for Proposals," May 13, 1997. The cover letter is signed by S. Dewey, J. Fowler, W. Graves, J. Jordan-Sweet, E. Kaplan, B. Ocko, and J. Shanklin.

50. John Shanklin to Clifford Krauss, email, May 19, 1997, in Early FOB, Summer 1997 notebook.

51. John Shanklin to Clifford Krauss, email, May 19, 1997, in Early FOB, Summer 1997 notebook.

52. FOB members also discovered that sarcasm did not help them get letters published. An outraged scientist wrote the *New York Times*, "It is a well known fact that the Long Island beaches are millions of times more radioactive than the worst leak ever at Brookhaven. Indeed, the total amount of uranium and thorium on these beaches measures in the tons, more than enough to fuel one huge reactor. This moderately radioactive material is a very real cause of cancer. What a magnificent contribution it would be to stop the millions of Federal money being spent patching the Long Island beaches. All sand movement should legally be subject to DOE radiation worker rules. Watch the politicians race to proclaim that all beaches should be close to protest us all." Jim Niederer, unpublished letter to the *New York Times*, June 13, 1997.

53. Frank Marotta to R. P. Crease, email, November 12, 2021.

54. John Shanklin to John Axe, August 25, 1997, BNL document.

55. Karl Grossman, "Suffolk Closeup," *East Hampton Star*, August 21, 1997, 1–16.

56. Karl Grossman "Suffolk Closeup," *East Hampton Star*, June 5, 1997, 1–12; Clifford Krauss, "Mounting Criticism Shadows Cleanup at Brookhaven," *New York Times*, June 2, 1997, B7.

57. Krauss, "Mounting Criticism Shadows Cleanup at Brookhaven."

58. John Rather, "High A-Levels in Water Spur Brookhaven Lab," *New York Times*, January 26, 1997.

59. Dan Barry and Andrew C. Revkin, "At 50 Brookhaven Lab Is Beset by Problems," *New York Times*, March 22, 1997.

60. Peter Maniscalco, Statement before Suffolk Legislature BNL Committee, November 5, 1998.

61. Eric Nelson, "Scarier than It Looks: The BNL Wars: Activist vs. Activist," *Long Island Voice*, April 16–22, 1998, 17.

62. Liam Pleven and Joe Haberstroh, "Lab Bolsters Damage Control," *Newsday*, April 12, 1997, A17. The newspaper also published how much the lab was paying to a consulting company. Joe Haberstroh, "Besieged Lab Hires PR Firm," *Newsday*, March 17, 1997, A4, A36. A related article is Joe Haberstroh and Liam Pleven, "Brookhaven Lab Seeks More PR in Top Post," *Newsday*, March 8, 1997, A4, A16.

63. Letter to the editor, *Newsday*, April 2, 1997.

64. Mona Rowe, "Labs' PR a Required Subject," *Newsday*, April 30, 1997.

65. Karl Grossman, "Suffolk Closeup," *East Hampton Star*, April 10, 1997.

66. Liam Plevin, "Activists: Pols Exploited Lab Leak," *Newsday*, June 18, 1997.

67. Nanette Essel, phone interview by R. P. Crease, July 7, 2021.

68. David Kamp, "The Tabloid Decade," *Vanity Fair*, February 1999.

69. Andrew Friedman, "Coke Fiends Cruise LIE in Limos," *Long Island Voice*, February 26–March 4, 1998.

70. Liam Pleven and Joe Haberstroh, "Coffee Mugs, but No Well," *Newsday*, April 3, 1997, A3, A36.

71. Alec Baldwin, letter to the editor, *East Hampton Star*, July 7, 1997. Some FOB members did not fail to point out that, due to a typo, the printed version of Baldwin's letter dates it as "997."

72. Andrew Lawler, "Meltdown on Long Island," *Science*, February 25, 2000, 1386. Baldwin subsequently walked back that remark; "Scientists at Brookhaven," letter to the editor, *Science*, April 7, 2000, 55.

73. Scott Cullen, phone interview by R. P. Crease, April 30, 2021.

74. Lauren Terrazzano, "Call It STAR Power," *Newsday*, December 16, 1998, A52.

75. Occurrence Fact Sheet, Brookhaven Medical Research Reactor (BMRR), June 3, 1997; Steve Centore, Nuclear Programs Division, BHG, "Multiple Personnel Contaminations Due to Airborne Contaminant," June 3, 1997.

76. "Five Workers Exposed to Radiation at BNL Medical Reactor; No Release of Radioactivity to Environment," BNL Press Release no. 97-54.

77. Liam Pleven, "Radiation Leak Deemed Harmless," *Newsday*, June 5, 1997.

78. Peter Bond, "Reactor Division Stand Down Meeting," notes, June 12, 1997.

79. "Media Update: Groundwater Contamination Near Underground Collection Tank," Brookhaven National Laboratory, June 11, 1997, BNL document.

80. Liam Pleven, "2nd Dangerous Isotope Cited in BNL Leak," *Newsday*, June 12, 1997, A27.

81. Emi Endo, "Brian Schneck, 30, Killed in Brookhaven Lab Accident," *Newsday*, June 22, 1997.

82. DOE News, "Statement by Secretary Federico F. Peña," June 20, 1997, R-97-055.

83. Office of Compliance, US Department of Energy, "Type A Accident Investigation Board Report on the June 20, 1997, Construction Fatality at the Brookhaven National Laboratory, Upton New York," https://www.energy.gov/sites/prod/files/2014/04/f14/9706bnl.pdf

84. Tara O'Toole to Peter Bond, June 27, 1997.

85. Tara O'Toole to Peter Bond, June 27, 1997.

86. "Construction Incident August 15 at Brookhaven Lab: No Worker Exposure, No Environmental Release Seen," BNL News Release no. 97-88.

87. For one of many calls to close the lab, see "Brookhaven Lab Debate: To Close or Not to Close," *Newsday*, April 8, 1997.

88. https://www.energy.gov/sites/prod/files/2018/10/f56/NNSA%20RAP%2060%20brochure%20-%20web.pdf.

89. Robert L. Park, "Tritium Exposure: Don't Touch That Flush Handle!," *What's New*, June 20, 1997, http://bobpark.physics.umd.edu/WN97/wn062097.html.

90. Paul Vitello, "Tritium's No Laughing Matter," *Newsday*, April 8, 1997, A8. The column began by referring to one of the author's articles about Brookhaven's early experience with community relations that mentioned a light-hearted musical review, which Vitello took out of context.

91. Dan Brown to Kara Villamil, September 6, 1997, BNL document.

92. Gregory Benford, *Cosm* (London: Aspect Press, 1998).

93. "As If BNL Didn't Have Enough Problems," *The Tide: The Voice of Brookhaven and Southampton Towns* 22, no. 11, 1997.

94. "The Truth about TWA-800: The UFO Controversy," website maintained by Kenny Young, a printout of http://www.home.fuse.net/task/TWA.htm, site no longer active.

95. Michael Colton, "Out There: They Thought UFOs Had Landed. A Case of Hysteria, Politics, Poison and Toothpaste," *Washington Post*, January 11, 1998, F1.

96. Agents X, Y and Z, "Brookhaven Lab: The Truth Is in There," *Hurricane Eye*, October 1997, p. 10.

97. Karl Grossman, "Nuclear Genocide: How Brookhaven National Laboratory Is Killing People," *Rage*, September 1997, 74–77.

98. Grossman, "Nuclear Genocide," 79. In September 1996 Smith announced a campaign to get alumni of the nine universities managing the laboratory to stop donating money to their university until they rid themselves of the lab that was "poisoning life on Long Island"; the effort was widely regarded as frivolous but was reported by *Newsday*. Smith's press release about it was released September 19 and the *Newsday* article on September 29. The next spring, Smith got his wish and the lab was taken out of the universities' hands, but his attacks on the lab were unrelenting.

99. Dan Rattiner, "The Lab 1947–2000: From Albert Einstein and Albert Schweitzer to a Plastic Rubble Cap," *Dan's Papers*, June 13, 1997, 27.

100. Dan Rattiner, "Babies & Baths: Shutting the Brookhaven Lab Is NOT an Option as Far as This Newspaper Is Concerned," *Dan's Papers*, June 20, 1997, 27.

101. Robert L. Park, "Radioactive Leaks: Hanford Waste Shows Up in Ground Water," *What's New*, November 28, 1997, http://bobpark.physics.umd.edu/WN97/wn112897.html.

102. Moncton had worked at Brookhaven in the mid-1970s, and done neutron scattering work for his PhD at the HFBR under Gen Shirane, the dean of the HFBR's user community. In the late 1970s, Moncton became interested in synchrotron radiation work and—with the NSLS still under construction—commuted from Setauket on Long Island to Stanford in California to do experiments at the light source there. In 1983 he began working on a plan to develop a yet more advanced facility, soon called the Advanced Photon Source (APS). In 1987, at the tender age of thirty-eight, he was chosen to head the billion-dollar APS project, still in the R&D stage.

103. Associated Universities, Inc., Executive Committee minutes, June 11–12, 1997.

104. Robert Park, "AUI Picks a Director for Brookhaven National Laboratory," *What's New*, June 13, 1997, https://web.archive.org/web/20161226173614/http://bobpark.physics.umd.edu/WN97/wn061397.html.

105. Liam Pleven, "Brookhaven Lab Reassured: No Talk of Closing, Says Congressman," *Newsday*, June 14, 1997, A19.

106. Lyle Schwartz, "BNL Employees," email, June 16, 1997.

107. Marsha Belford, "Subject to DOE Approval, AUI Votes Moncton Director," *Brookhaven Bulletin*, June 20, 1997, 1, 3.

108. Peter Bond, "A Message to Participants of the Brookhaven National Lab Clinical Research Center and Radiation Therapy Facility," undated flyer, BNL document.

109. "EPA Takes Look at Brookhaven Laboratory," DOE Press Release, July 15, 1997.

110. Marsha Belford, "Congressman Michael Forbes Apologizes for 'Hurtful' Words; Advocates BNLers' Continued Employment, Lab's Science Mission," *Brookhaven Bulletin*, July 18, 1997, 1, 3.

111. BNL Press Release no. 97-74, https://www.bnl.gov/bnlweb/pubaf/pr/1997/bnlpr091097 .html.

112. Peter Bond, notes from meeting.

113. Gary Schroeder, "Elevated Tritium Level at the Sewage Treatment Plant, July 1997," Environmental Protection Office, Brookhaven National Laboratory. From its typical level of 2,000 picocuries per liter the amount had suddenly flared to 90,000 picocuries per liter at the entrance of the treatment plan and 67,000 picocuries per liter at the outfall. The tritium did not get off-site.

114. These included facilities such as the Alternating Gradient Synchrotron, National Synchrotron Light Source, Brookhaven Linac Isotope Production Facility, Brookhaven Medical Research Reactor, and the closed HFBR; the Physics, Chemistry, Biology and Medical Departments; the Departments of Applied Science and Advanced Technology; and the hazardous waste management facility.

115. Liam Pleven, "New Tritium Problems at Brookhaven Laboratory," *Newsday*, July 30, 1997, A21.

116. John Stringer, Chair, Basic Energy Sciences Advisory Committee, to Martha Krebs, November 22, 1997.

117. Michael Forbes to Secretary Peña, August 29, 1997.

CHAPTER 4

1. David Brower with Steve Chapple, *Let the Mountains Talk, Let the Rivers Run: A Call to Those Who Would Save the Earth* (New York: HarperCollins, 1995), 27.

2. Michael Forbes and Alfonse D'Amato, "Media Advisory," September 1, 1997.

3. Liam Pleven, "Pols Want LI Reactor Shut Down," *Newsday*, September 2, 1997.

4. Bruce Lambert, "D'Amato Bill Would Shutter Lab's Reactor," *New York Times*, September 3, 1997.

5. Michael Forbes and Alfonse D'Amato, "D'Amato, Forbes: Brookhaven Lab's Damaged Nuclear Reactor Must be Permanently Shut Down," press release, September 2, 1997.

6. Marsha Belford, "D'Amato and Forbes Call for Permanent Shutdown of HFBR; They Push Legislation for Reactor's Decommissioning," *Brookhaven Bulletin*, September 5, 1997, 2.

7. Belford, "D'Amato and Forbes Call for Permanent Shutdown of HFBR."

8. Belford, "D'Amato and Forbes Call for Permanent Shutdown of HFBR."

9. S. 1140, "To Prohibit Reactivation of the High Flux Beam Reactor at Brookhaven National Laboratory," introduced by Mr. D'Amato, September 2, 1997. The bill was not acted on, but Forbes and D'Amato continued to try to use legislation to force the HFBR shutdown, adding an amendment to an appropriation bill to forbid spending money to restart of HFBR for the next fiscal year, FY 1998 (October 1997–1998). On October 13, H.R. 2203, Section 512 of

the FY 1998 Energy and Water Development Appropriations Bill, which stated, "None of the funds made available in this or another Act may be used to restart the High Flux Beam Reactor," was signed into law. It was symbolic, for Secretary Peña had indicated that he would not make a decision until several months into 1998, and DOE officials said they did not plan to reopen the reactor in any case in the next year. Nothing appeared to prohibit restart after September 1998, though this was hotly disputed by Forbes and D'Amato. "They can put any spin on it they want," Forbes said, "but the fact is that this puts us on the path to a permanent shutdown." After lab chemist Steven Dewey wrote a letter of protest, Forbes heatedly defended his decision, handwriting at the end of his letter of reply, "Your management betrayed you—no matter how much you want to blame politicians—they ignored their responsibilities!" Michael P. Forbes to Stephen L. Dewey, September 22, 1997.

10. "Political Cave-In: Pressing to Shut Brookhaven's Big Reactor Is Premature and Potentially Harmful to LI," *Newsday*, September 4, 1997, A46.

11. D. Allan Bromley to Alfonse D'Amato, reprinted in *American Institute of Physics Bulletin of Science Policy News* 112 (September 16, 1997), https://www.aip.org/fyi/1997/bromley-letter-legislation-close-high-flux-beam-reactor.

12. "Statement by Dr. Martha Krebs, Director, Office of Energy Research," L-97-092, September 2, 1997.

13. Robert L. Park, "Brookhaven: Et Tu Brute?," *What's New*, September 5, 1997, http://bobpark.physics.umd.edu/WN97/wn090597.html.

14. The Friends of Brookhaven were joined by two other more recently formed groups, the Brookhaven Scientists Association and the newly formed Brookhaven Retired Employees Association.

15. FOB, handwritten list of signs and comments, undated.

16. "Petition to Representative Forbes and Senator D'Amato," Friends of Brookhaven, Brookhaven Science Association, and the BNL Retired Employees Association, September 1997.

17. The address was 1500 William Floyd Parkway, Shirley, NY 11967.

18. Marsha Belford, "500 BNLers Rally for HFBR at Forbes's Local Office, Demanding 'Science Fact, Not Science Fiction,'" *Brookhaven Bulletin*, September 12, 1997, 2.

19. Belford, "500 BNLers Rally for HFBR."

20. Belford, "500 BNLers Rally for HFBR"; "Rally in Support of BNL," flyer dated September 4.

21. Federico Peña, letter to Alfonse D'Amato and Michael Forbes, September 5, 1997.

22. Peter Bond, "M. Krebs 9/4/97, Debrief D'Amato meeting," notes, September 4, 1997.

23. Marsha Belford, "Forbes Will Not Change His Position on the HFBR Under 'Any Condition,'" *Brookhaven Bulletin*, September 12, 1997, 3.

24. D. Ports to A. Queirolo, "Report of Elevated Sanitary Tritium Levels Observed at the HFBR," October 3, 1997. On August 27, routine sampling of the wastewater from the sinks, toilets, and

floor drains had shown an unusually high tritium concentration of 49,300 picocuries per liter, which was above the 1997 daily average of 10,100 picocuries per liter. It turned out to be due to a change in the way that condensate from an air filter had been disposed of. The temporary increase resulted in a tritium concentration of 5,000 picocuries per liter on August 28 at the sewage treatment plant, well below the EPA's drinking water standard of 20,000 picocuries per liter.

25. Belford, "Forbes Will Not Change His Position on the HFBR."

26. Michael Forbes, letter to constituents, September 8, 1997.

27. Jennifer O'Connor, letter to the editor, *Three Village Herald*, October 15, 1997, 13.

28. "Write Better Letters," *Brookhaven Bulletin*, September 12, 1997, 4.

29. Shelter Island Police Department, "Arrest," press release, September 11, 1997.

30. Liam Pleven, "Police: BNL Worker Threatened Critic," *Newsday*, September 14, 1997.

31. "Police Blotter," *Shelter Island Reporter*, September 4, 1997, 6.

32. "1,200 BNLers Respond to Forbes's 'Letter of Condemnation' with Calls for Truth, Due Process, 'the American Way," *Brookhaven Bulletin*, September 26, 1997, 1.

33. John Shanklin, phone interview by R. P. Crease, January 9, 2021.

34. Karl Grossman, "Suffolk Closeup: Arrogant Stance," *Southampton Press*, October 2, 1997.

35. Peter Bond, "Response to Congressman Michael Forbes's Letter," memo to BNL employees, September 19, 1997.

36. Associated Universities, Inc., "A Public Response to Congressman Forbes from Associated Universities, Inc.," *Newsday*, October 1, 1997, B9. Along with this response, AUI continued to involve itself vigorously in the lab, making enormous progress in the cleanup of contamination and organizing external reviews of the lab's scientific programs.

37. As an example, Alfonse M. D'Amato, letter to the editor, *Newsday*, September 29, 1997.

38. Peter Bond and Mike Forbes, "At Issue: The Future of the HFBR," Channel 12 News, September 14, 1997.

39. Long Island Association, "LIA: Let U.S. Department of Energy Decide Fate of BNL Reactor; Directors Unanimously Support Committee Review Process," press release, September 15, 1997.

40. "ABCO/Community Groups Press Conference September 17, 1997 at BNL Main Gate," flyer.

41. The program for the event: Hillel Foundation for Jewish Life, State University of New York at Stony Brook, "High Holiday Services 5758–1997," September 1997.

42. S. Adler et al. (E787 Collaboration), "Evidence for the Decay $K^+ \rightarrow \pi^+ \nu \bar{\nu}$," *Physics Review Letters* 79, no. 2204 (September 22, 1997).

43. John Stringer to Robert Birgeneau, April 25, 1997.

44. Synchrotron Radiation Light Source Working Group, "Report of the Basic Energy Sciences Advisory Committee," October 8–9, 1997.

45. Robert Birgeneau, "Report of the Basic Energy Sciences Advisory Committee Panel on D.O.E. Synchrotron Radiation Sources and Science," November 1997, https://science.osti .gov/-/media/bes/besac/pdf/Sync_panel.pdf.

46. "Interim Report of the BNL Facility Review: Review of Potential Environmental Release Points," September 9, 1997.

47. W. J. Brynda to J. Carelli, "Telephone Conversation with R. W. Powell on 9/30/97 Regarding Investigation into Source of Water in BGRR Air Exit Ducts," October 1, 1997, BNL document.

48. "Secretary Regrets Resignation of Dr. Tara O'Toole, Assistant Secretary for Environment, Safety and Health," DOE Press Release R-97-093, September 18, 1997.

49. George Lobsenz, "Get Involved . . . or Pay the Price, O'Toole Tells Labs," *Energy Daily*, October 9, 1997, 1.

50. Peter Bond to Tara O'Toole, September 24, 1997.

51. Tara O'Toole to Peter Bond, September 26, 1997.

52. Peter Bond to Tara O'Toole, September 29, 1997.

53. Tara O'Toole to Peter Bond, October 2, 1997.

54. Tara O'Toole, phone interview by R. P. Crease, September 14, 2020.

CHAPTER 5

1. The actual fee was determined by DOE grades on performance in different areas, such as science and ES&H, and was never the maximum.

2. The formal decision of the winning bid was to be made by Franklin Peters, DOE's Acting Associate Deputy Secretary for Field Management, although most members of the evaluation team seem to have been unaware of that.

3. Steven Silbergleid, phone interview by R. P. Crease, December 14, 2020.

4. Steven Silbergleid, phone interview by R. P. Crease, December 14, 2020.

5. "AUI Won't Bid to Manage BNL," August 5, 1997.

6. Lyle Schwartz, "Statement of Lyle H. Schwartz, AUI President," August 5, 1997.

7. "Dismissed Manager at Lab Calls Energy Dept. Unfair," *New York Times*, August 6, 1997.

8. In 1965, when General Electric decided to stop managing the Hanford Laboratory in Washington State, and the AEC split the lab's activities into an R&D piece and a cleanup piece, Battelle won the competition for the R&D piece, and renamed it Pacific Northwest Laboratory (PNL). The lab did not have user facilities and would not become a national lab until 1995 when it was renamed Pacific Northwest National Laboratory (PNNL).

9. Kevin Mayhood, "Battelle's World: Columbus-Based Research Giant Extends Its Global Reach," *Columbus Dispatch*, January 25, 2009.

10. Mona Rowe, phone interview by R. P. Crease, September 20, 2020.

11. In 1961, New York State acquired land in Upstate New York for a Western New York Nuclear Service Center (WNYNSC). A year later, New York State leased the WNYNSC to the Davison Chemical Company, to create a reprocessing facility called Nuclear Fuels Services, Inc. (NFS). The NFS served as a nuclear fuel reprocessing center from 1966 to 1975. During this time the NFS stored waste both from commercial nuclear power plants and from reprocessing. This included 660,000 gallons of highly radioactive liquid waste stored in an underground tank. In 1980, the NFS decided that its reprocessing center was not financially rewarding, and ceded the facility with its wastes back to New York State. New York State then initiated discussions with the Energy Research and Development Administration (ERDA), the DOE's predecessor agency, regarding how to handle the environmental restoration.

12. Venkatesh Narayanamurti, phone interview by R. P. Crease, February 20, 2020.

13. Rochard Lahey, phone interview by R. P. Crease, June 1998.

14. DOE News, "DOE Extends Deadline for Responding to Request for Proposals for Brookhaven Contract," August 27, 1997.

15. Excerpt from Jack M. Wilson Memoirs (unpublished), pp. 153–161, used with permission.

16. "Bidder for Brookhaven Centre Bails Out," *Nature*, September 11, 1997.

17. John H. Marburger III, *Science Policy Up Close*, ed. Robert P. Crease (Cambridge, MA: Harvard University Press, 2015).

18. On September 5, the BSA team trucked its proposal—sixteen boxes of documents, one copy for each SEB member and other officials—from its proposal center in Columbus to the Department of Energy's Chicago office. "In case of accident," Madia said, "you have a whole backup, so there's another 16 boxes in reserve along with another truck. You can't spend millions of dollars on these things and have a stupid truck accident get in the way. Later, we had a backup private plane. Nowadays, of course, proposals are uploaded."

19. "Energy Department Extends Deadline for Brookhaven Proposals," DOE News, September 8, 1997; Amendment S007.

20. Liam Pleven, "Bidder Pill to Swallow," *Newsday*, September 9, 1997.

21. A. Nonymous, "711 and Mattel to Run BNL," September 5, 1997.

22. IITRI was so often assumed to be part of the Illinois Institute of Technology that its press releases concluded with the following "Editor's Note: IITRI's full corporate name is IIT Research Institute. Please do not use 'Illinois Institute of Technology Research Institute.'"

23. BSA's upper management included two people from Stony Brook, Director Marburger and Deputy Director for Science Peter Paul, and five people from Battelle—Thomas Sheridan (deputy director of operations), Ken Brog (associate director for ESH), Adrian Roberts (associate director of applied research and technology), Michael Schlender (associate director of environmental cleanup), and Gregory Fess (general counsel). Both Peter Paul and Robert Birgeneau (who was not present in Columbus) had resigned as AUI trustees as they were now

part of the BSA management structure; Paul was the designated deputy director for science and Birgeneau was the MIT representative on the board.

24. Steven Silbergleid to Terry Buckner, fax, October 7, 1997.

25. As Secretary Peña put it in the congressional notification letter, "Section 301(a) of the recently enacted Energy and Water Development Appropriations Act, 1998 (Pub. Law 105-62) prohibits the use of certain appropriations to award a management and operating contract unless such contract is awarded using 'competitive procedures' or the Secretary of Energy grants a 'waiver' to the restriction. Section 301(b) requires that the Secretary provide a report notifying the Subcommittees on Energy and Water Development, of the Committees on Appropriations of the House of Representatives and the Senate, of the 'waiver' at latest 60 days before contract award. The purpose of this letter is to provide such notice in anticipation of the award of a contract for the management and operation of the Brookhaven National Laboratory (BNL), a federally funded research and development center (FFRDC)." Federico Peña to Joseph M. McDade, Chairman, Subcommittee on Energy and Water Development, Committee on Appropriations, November 5, 1997.

26. J. H. Marburger to R. P. Crease, email, November 10, 1997, SBU SC.

27. Brookhaven Science Associates, "Battelle, SUNY-Stony Brook Team Wins Brookhaven Contract," Press Release 63-97.

28. "Introducing: Brookhaven Science Associates (BSA)—New Management Contractor for BN," undated but printed before November 24, 1997. BSA would have a 16-member board of directors, five appointed by SBU, five by Battelle, and one each from the six universities: Columbia, Cornell, Harvard, Massachusetts Institute of Technology, Princeton, and Yale. SBU President Shirley Kenny would chair the BSA board for the first two years, with Battelle's CEO Douglas Olesen the vice-chair.

29. Franklin Peters, "Selection Statement," November 24, 1997.

30. The evaluation document praised both bidders; a key factor for the choice of BSA was "its clear superiority compared to IITRI–Westinghouse in the area of Scientific and Technological Programs," which was the highest weighted of all evaluation criteria. This was largely based on their having more experience in the areas of BNL's science mission. In the areas of operations IITRI–Westinghouse was viewed as being stronger in a number of areas, including the all-important ES&H; Westinghouse's corporate experience indicates "a broad scope of recognized ES&H leadership and record of accomplishment at DOE facilities." Both teams showed they could effect the "necessary cultural change," but BSA had more experience in relevant research areas. IITRI had no experience in nuclear physics, and only limited experience in high-energy physics. BSA therefore was rated as "outstanding" in scientific programs management, with IITRI Brookhaven "satisfactory." Both were said to have "outstanding" proposed lab directors. "For the reasons set forth above, it is my determination that the selection of Brookhaven Science Associates provides the best overall value to the government." Peters, "Selection Statement."

31. "Statement from Secretary of Energy Federico Peña on the Announcement of a New Contractor for Brookhaven National Laboratory," in "Energy Department Announces New Contractor to Improve Management of Brookhaven Lab," DOE Press Release R-97-130, November 25, 1997.

CHAPTER 6

1. Jack Sirica, "SUNY Sees Lab as Aid to School," *Newsday*, November 25, 1997. Princeton, Stanford, and Iowa State all ran facilities for the DOE Office of Science that were not then designated as national laboratories.

2. Charlie Zehren and Liam Pleven, "'Promising' Beginning: Stony Brook Team Wins Contract to Run Troubled Brookhaven Lab," *Newsday*, November 26, 1997, A38.

3. Sirica, "SUNY Sees Lab as Aid to School."

4. Robert L. Park, "Radioactive Leaks: Hanford Waste Shows Up in Ground Water," *What's New*, November 28, 1997, http://bobpark.physics.umd.edu/WN97/wn112897.html.

5. Charlie Zehren, "He's Ready to Steer: But Brookhaven's New Director Faces Uncertain Path," *Newsday*, November 30, 1997, A6, A43.

6. John Marburger, draft letter to the editor *Newsday*, undated. *Newsday* did not publish it.

7. Spencer Rumsey, "'It Takes More than Science to Do Science,' the *Newsday* Interview with John Marburger," *Newsday*, December 30, 1997, A29–A30.

8. John H. Marburger III, "Letter to 'Franklin' about the High Flux Beam Reactor," in Marburger, *Science Policy Up Close*, ed. Robert P. Crease (Cambridge, MA: Harvard University Press, 2015), 92–105.

9. John T. McQuiston, "Branch of State University Will Run Brookhaven Lab," *New York Times*, November 26, 1997, B5; Karl Grossman, "B.N.L. Gets New Managers," *East Hampton Star*, December 4, 1997.

10. "Stony Brookhaven," editorial, *Newsday*, November 26, 1997, A46.

11. Michael Gross, "See Alec Run: Tired of Hollywood, Alec Baldwin Is Preparing for the Role of His Career: Candidate," *New York Magazine*, November 24, 1997.

12. *Mirabella*, January–February 1998, 66.

13. Stephen Glass, "The Boys on the Bus: Alec Baldwin's New Role: The Candidate," *New Republic*, December 8, 1997, 25–29.

14. "Feral Democrat," editorial, *Newsday*, December 13, 1997, A24.

15. Lauren Terrazzano, "Call It STAR Power," *Newsday*, December 16, 1998, A7, A52–A53.

16. Valerie Cotsalas, "Fire at Brookhaven Lab Traced to Hay in Debris," *New York Times*, April 9, 2000.

17. Karl Grossman, "Suffolk Closeup," *East Hampton Star*, October 23, 1997, 1–14.

18. David Friedson to Federico Peña, November 24, 1997.

19. "LIPC's Community Activity at the Brookhaven Lab," *Long Island Progressive*, Spring–Summer 1998, 27.

20. Erik Nelson, "Scarier than It Looks. The BNL Wars: Activist vs. Activist," *Long Island Voice*, April 16–22, 1998, 16.

21. Andrew Lawler, "Meltdown on Long Island," *Science*, February 25, 2000, 1384.

22. Bill Smith, phone interview by R. P. Crease, June 15, 2021.

23. Fenton Communications, "Alec Baldwin to Host Star Fundraiser," advisory for October 14, 1997.

24. Liam Pleven, "New Pressure on Lab: Critics Mobilize for a Fight," *Newsday*, October 18, 1997, A3.

25. Liam Pleven, "New Pressure on Lab: Critics Mobilize for a Fight," *Newsday*, October 18, 1997, A3.

26. In addition to correcting facts in the report, the discussion focused on the current status, future plans, and discussion of the report's recommendations. There was substantial agreement with the latter, in particular: (1) adopt cutting-edge environmental management practices and systems, (2) do not restart the HFBR before corrective measures and systems were in place to prevent leaks, and (3) increase outreach activities.

27. Alec Baldwin to Peter Bond, October 27, 1997.

28. US General Accounting Office, "Information on the Tritium Leak and Contractor Dismissal at the Brookhaven National Laboratory," GAO/RCED-98-26, November 1998, 29.

29. Quoted in Liam Pleven, "Water Near Lab Deemed Safe," *Newsday*, October 11, 1997, A8.

30. Quoted in Liam Pleven, "Water Near Lab Deemed Safe," *Newsday*, October 11, 1997, A8.

31. "Department of Energy Workers Cancer Benefit Program," National Cancer Benefits Center, https://www.cancerbenefits.com/cancer-benefit-programs/dept-of-energy-workers/.

32. Associated Universities, Inc., Board of Trustees Meeting minutes, April 16–17, October 22–23, 1997.

33. Roger Zureck to Federico Peña, letter, December 18, 1997.

34. David Carney, "Letter to the Editor," *Southampton Press*, October 2, 1997.

35. John Shanklin to Alec Baldwin, October 17, 1997.

36. Ed Kaplan, phone interview by R. P. Crease, October 1, 2020.

37. John Shanklin, phone interview by R. P. Crease, January 9, 2021.

38. "FOB," Shanklin notebook, 1997.

39. Ed Castner to anderson@wshu.org, email, November 5, 1997; WSHU Underwriting Copy Sheet.

40. Liz Seubert, "Community Supports BNL & Science: 18,285 Sign Petition," *Brookhaven Bulletin*, January 9, 1998, 2.

41. Dan Oldham, "Petition Deadline," Dan Oldham to Lablist2, November 6, 1997.

42. Jerry Cimisi, "The Fixer," *Dan's Papers*, January 30, 1998, 15.

43. Friends of Brookhaven, "Moynihan Call-In Day Friday, October 31," flyer.

44. John Shanklin, phone interview by R. P. Crease, January 9, 2021.

45. Jay M. Gould, *The Enemy Within: The High Cost of Living Near Nuclear Reactors* (New York: Four Walls Eight Windows, 1996), 11. One review said that Gould lacked "even a rudimentary understanding of the fields of radiation biology and epidemiology," that he had "violated many of the principles of good science," and that knowledgeable people should "point out the lack of any scientific foundation whatsoever for the claims that are made." Dade W. Moeller and Steven E. Merwin, "Misinformation on the Effects of Low-Level Radiation—Commentary on the Book *The Enemy Within*," *Health Physics Society Newsletter*, October 1996, 15–16.

46. Quoted in Karl Grossman, "Suffolk Closeup: Another Dose," *Southampton Press*, April 6, 1994.

47. Concerned Brookhaven National Laboratory Scientists, "Comments on 'The Long Island Breast Cancer Epidemic': Evidence for a Relation to the Releases of Hazardous Nuclear Wastes," Center for Management Analysis Occasional Papers, December 1994. See also Philip Boffey, "Ernest J. Sternglass: Controversial Prophet of Doom," *Science*, October 10, 1969, 195–200.

48. Peter Bond to William Weida, email, December 7, 1997, BNL document.

49. Ron Stanchfield to William Weida, email, December 11, 1997.

50. Peter Bond to Scott Cullen, letter, December 30, 1997, BNL document.

51. William J. Weida to Federico Peña, January 12, 1998.

52. Niraj Warikoo, "Protest at Lab Leads to 8 Arrests," *Newsday*, December 7, 1997, A21.

53. Karl Grossman, "Suffolk Closeup," *East Hampton Star*, December 25, 1997.

54. Charlie Zehren, "Bishop Leads Push against Lab," *Newsday*, December 18, 1997, A37.

55. Transcription of a recording, December 17.

56. Zehren, "Bishop Leads Push against Lab."

57. Quoted in Robert Scheuerer, *Newsday*, January 1, 1998.

58. Doug Ports, "Balancing Science and the Environment," unpublished handout, 1997. US Catholic Conference, 60. *Renewing the Earth* (USCCB Publishing, 1992). The US Catholic Conference took part, issuing its own monograph for Earth Day, entitled "Renewing the Face of the Earth."

59. Peter Strugatz to Peter Bond, December 10, 1997.

60. Robert Weimer, letter to the editor, *Newsday*, December 28, 1997, B3.

61. Scott Cullen, "Not Ready for Media Time," *Newsday*, January 2, 1998, A38.

62. John H. Marburger III, interview by R. P. Crease, May 2011.

63. Jordan Rau, "Actor without an Audience: Baldwin Bars Reporter from Lab Meeting," *Newsday*, December 23, 1997.

64. John H. Marburger III, "Notebook: October 1997–February 1998," December 23, 1997.

65. John H. Marburger III, interview by R. P. Crease, Stony Brook University, May 2011.

66. Anita Cohen, "Public Workshop Thursday on a Potential Community Advisory Council," acohen to BNL Labwide Broadcasts, November 17, 1997, BNL document.

67. Gina Kolada, "Epidemic That Wasn't," *New York Times*, August 29, 2002. Other national laboratories were also accused of being centers of higher than normal levels of cancer; for Oak Ridge, see Michael Fumento, "A Newspaper Invents a Nuclear Health Scare," *Wall Street Journal*, November 12, 1998.

68. According to the Suffolk County Task Force report, the rate was 4.1 cases of rhabdomyosarcoma annually per million people under age 19 in Suffolk County, 5.6 in Nassau, 6.4 in Queens, 7.0 in Brooklyn, and 5.3 throughout New York State.

69. The incidence of breast cancer was lower in the area immediately surrounding the lab than farther east, in Long Island's North and South Forks. The Task Force found that, overall, there was no higher incidence of any kind of cancer in areas surrounding the lab. The report did, however, find a higher incidence of cancer east of the lab, which had many farms whose properties had been frequently sprayed with pesticides.

70. Roger Grimson and Dawn Triche [a member of the Community Work Group], "Report to the Suffolk County Legislature from the Brookhaven National Laboratory Environmental Task Force: Evaluation of Allegations of Contamination and Risk Associated with the Operation of Brookhaven National Laboratory, Part I: Epidemiology," Executive Summary, January 26, 1998.

71. Its members were drawn from surrounding neighborhoods and included scientists, engineers, New York and Suffolk County health officials, and antinuclear activists. The committee began working in October 1996, and would issue its final report in January 1998. It stated the following: "The conclusions of the epidemiological cancer study are that (1) cancer rates of all types of cancers studied are not elevated near BNL and (2) there is no evidence that rates among the four sectors are significantly different from each other or are correlated with underground plume or wind directions. Also, there is no evidence that childhood rhabdomyosarcoma incidence is elevated in Suffolk County or in the [15-mile] circle encompassing BNL during the study period 1979–1983 for which the Registry rhabdomyosarcoma data were available at the time of the request." Grimson and Triche, "Report to the Suffolk County Legislature from the Brookhaven National Laboratory Environmental Task Force."

72. Undated letter from EPA to Robert Hughes, received on March 4, 1998.

73. Three violations were cited: a failure to follow procedures when moving dummy fuel elements within the HFBR, not following procedures while irradiating experimental samples

at the BMRR, and having an uncertified radiation control technician. According to the DOE press release, the violations reflected "a trend of non-compliance with regulatory and Brookhaven procedural requirements." "Energy Department Cites Associated Universities for Safety Violations," DOE News, December 22, 1997, R-97-142.

74. Robert L. Park, "Brookhaven: DOE Sends the Lab a Christmas Greeting," *What's New*, December 26, 1997, http://bobpark.physics.umd.edu/WN97/index.html.

75. Public Affairs Office, "Brookhaven Lab Spent Almost $33 Million on LI in 1997," Press Release no. 97-126.

76. Peter Bond, memo to employees, email, December 31, 1997.

77. Niraj Warikoo, "Prayers Held to Protest Lab," *Newsday*, December 22, 1997, A26.

78. Peter Maniscalco to Dean Helms, December 3, 1997, BNL document.

CHAPTER 7

1. That prompted one retired military officer to write Williams a letter calling him "vile" and "corrupt" and accusing him of being "a stooge and at worst a co-conspirator in [the psychic's] dishonest and cruel violation of so many families while chasing so many dollars." The letter ended, "[H]ave you lost your honor?" Lt Colonel Hal Bidlack, PhD, USAF (Ret.), "An Open Letter to Lt. Commander Montel Williams," http://stopsylvia.com/articles /openlettertomontel.

2. Scott Cullen to R. P. Crease, email, September 1, 2021.

3. "Program Appearance Agreement Revised," *The Montel Williams Show*, August 22, 1997.

4. Statistics from the New York State Cancer Registry showed that the incidence of rhabdomyosarcoma in Suffolk County was not the highest among New York State counties, with Nassau ranking in the top five, and Suffolk—where the lab is located—not even in the top ten.

5. American Cancer Society, "Rhabdomyosarcoma Causes, Risk Factors, and Prevention," December 14, 2021, flyer and web page, https://www.cancer.org/content/dam/CRC/PDF /Public/8804.00.pdf. References cited include M. F. Okcu and J. Hicks, "Rhabdomyosarcoma in Childhood and Adolescence: Epidemiology, Pathology, and Molecular Pathogenesis," UpToDate, https://www.uptodate.com/contents/rhabdomyosarcoma-in-childhood -and-adolescenceepidemiology-pathology-and-molecular-pathogenesis (accessed May 21, 2018); L. H. Wexler, S. X. Skapek, and L. J. Helman, "Chapter 31: Rhabdomyosarcoma," in *Principles and Practice of Pediatric Oncology*, 7th ed., ed. P. A. Pizzo and D. G. Poplack (Philadelphia, PA: Lippincott Williams & Wilkins; 2016).

6. Dan Oldham to Lablist2, January 8, 1998, BNL document.

7. Kara Villamil to Kenneth T. Smith, February 11, 1998, BNL document.

8. Richard B. Setlow, letter to the editor, *Southampton Press*, February 19, 1998.

9. Dan Oldham to Lablist2, "SmartAlec," January 11, 1998, BNL document.

10. Dan Oldham to Lablist2, "SmartAlec," January 11, 1998, BNL document.

11. Dan Oldham, "FOB Letter Writing Canceled," email to "Lablist 2," January 12, 1998, BNL document.

12. "Don't Confuse Me with the Facts," unsigned editorial, *The Village Times*, January 15, 1998, 46.

13. James S. Frank, letter to the editor, *Three Village Times*, January 29, 1998, 42.

14. Charlie Zehren, "'Montel' Takes on Brookhaven Lab," *Newsday*, January 9, 1998, A33.

15. Stephen Dewey to Brian Cullen, January 30, 1998.

16. Stephen L. Dewey to Mona Rowe, "Yet Another Letter . . . ," January 27, 1998.

17. Jan Burman, "Stand Up for the Brookhaven Lab," *Newsday*, January 9, 1998.

18. Leon Jaroff, letter to the editor, *East Hampton Star*, June 18, 1998.

19. Leon Jaroff, letter to the editor, *Dan's Papers*, January 23, 1998, 31.

20. Leon Jaroff, letter to the editor, *East Hampton Star*, February 5, 2000.

21. Otto G. Raabe, letter to the editor, *New York Times*, January 22, 1998.

22. For rates of incidence of rhabdomyosarcoma, see chapter 6, note 68.

23. For rates of incidence of breast cancer, see chapter 6, note 69.

24. Here's the conclusion of the ATSDR study, in the final release of its Public Health Assessment: "ATSDR concludes the BNL site does not currently pose a health hazard. The site contains no current, completed exposure pathway(s) to chemicals or to radionuclides at levels of public health concern. Past airborne radionuclide releases from the various reactors were apparently large, but the radiological doses were relatively small. The reason for the low doses is the properties of the respective radionuclides released. These radionuclides were either low energy emitters (tritium) or were not absorbed to a great extent in the body (argon-41). Past exposures to groundwater contaminants do not appear to have been at levels that would result in adverse health effects. Before sampling, ATSDR modeled the groundwater contaminant plumes to determine the likelihood of well contamination. Our modeling of the groundwater plumes indicated that the levels of contaminants in the residential wells would not have been above those detected at the time of sampling." US Department of Health and Human Services, Public Health Service, Agency for Toxic Substances and Disease Registry, July 6, 2011, xiv, https://www.bnl.gov/gpg/files/Misc_reports/BrookhavenNatlLabFinalPHA08012011.pdf.

25. Janine Giordano, "Cancer Study Flawed," *Suffolk Life*, March 18, 1998, 1.

26. Pete Maniscalco, letter to the editor, *Newsday*, January 19, 1998.

27. Charlie Zehren, "No Cancer Link Found," *Newsday*, January 23, 1998.

28. DOE News, "NY State Cancer Registry to Look at Cancer Rates among 21,000 Past and Current BNL Workers," January 29, 1998.

29. The Energy Employees Occupational Illness Compensation Fund law (commonly known as the Atomic Workers Bill) passed by Congress in 2000 was primarily focused on covering costs of health impacts of workers who were involved in nuclear weapons production or processing at DOE facilities and industries. However, the bill was much broader than the nuclear weapons cohort as all DOE facilities, including national laboratories, were included. The primary health issues were cancers known to be caused by radiation and issues of beryllium exposure. The fact that BNL was on the very long list of facilities was unrelated to 1997 or to the push for an employee health study. The criteria for eligibility were established for each organization (type of cancers, length of employment, exposure records). For example, BNL records of radiation exposure in earlier years (1947–1979) were not complete so certain of its employees were included in the evolving criteria (as were those at other labs). The program continues today and is run through the Department of Labor.

30. Jordon Rau, "Brookhaven Lab-Cancer Study Set," *Newsday*, January 30, 1998, A35.

31. Joseph H. Baier, P.E., Director of Environmental Quality, to William E. Gunther, "Tritium Analyses—Public Drinking Water Supply," February 10, 1998, BNL document.

32. Elizabeth Kolbert, "The Greening of D'Amato, up from Zero," *New York Times*, February 5, 1998, B1.

33. Erik Nelson, "BNL Plays Little Boy to Fat Man Hanford," *Long Island Voice*, February 19–25, 1998, 9.

34. "Memorandum to All BNL Employees," from the AUI Board of Trustees, Officers and Staff, *Brookhaven Bulletin*, February 27, 1998, 1.

35. Bill Gunther, "Media Attention on 'New' Environmental Contamination Issue," in Kara Villamil to lab employees, March 2, 1998, BNL document.

36. John H. Marburger III, "Notebook: February 1998–May 1998," February 28, 1998.

37. John H. Marburger III, "Notebook: February 1998–May 1998," March 1, 1998.

38. Associated Press, "Water Wasn't Imperiled Brookhaven Chief Says," *New York Times*, March 3, 1998, B6.

39. John H. Marburger III, "Notebook: February 1998–May 1998," March 1, 1998.

40. John H. Marburger III, "Notebook: February 1998–May 1998," March 1, 1998.

41. STAR Foundation, "Undisclosed Radiation Threat Emanating from Brookhaven National Laboratory," press release, no date.

42. Scott Cullen, phone interview by R. P. Crease, April 30, 2021.

43. Leon Jaroff, letter to the editor, *East Hampton Star*, March 12, 1998.

44. Joanna Fowler, STAR Fundraiser Notes, March 3, 1998.

45. Jeff Johnson, "Course Change for Brookhaven?," *Chemical and Engineering News*, January 5, 1998.

46. In the beginning, the management arrangement for national laboratories was known as GOCO, for government owned, contractor operated. In it, the government would provide funding for the laboratories, but interposed between them was a nonprofit buffer institution that enabled the laboratories to be run like a university—with the flexibility to find researchers, to pursue missions, and to pivot—rather than an agency directly connected to the federal government, such as the Tennessee Valley Authority. In 1947 Brookhaven's management involved a tripod of parties—BNL, AUI, the AEC; the lab, the buffer, the government—acting in close partnership and focused on promoting science for the national good. In 1974–1977, the governmental leg became more inflexible as the AEC morphed first into ERDA and then into the DOE, a much larger and more detached organization in which the national labs were only small items in its portfolio. Ever since, the GOCO arrangement had been slowly eroded as the federal government imposed increasing strictures on the contractors, particularly with respect to liability. The lab contracting system had not become GOGO—government owned, government operated—that is, national labs were not operated like the Tennessee Valley Authority or like the service labs such as the Naval Research Lab in the Defense Department, or not yet. But the contract between the DOE and BSA signed in January 1998 was a significant step in that direction. On the beginnings of the federal contracting system for science, see Harold Orlans, *Contracting for Atoms* (Washington, DC: Brookings Institution, 1967).

47. Charles Zehren, "New Chemistry," *Newsday*, March 2, 1998, A5.

48. "Adverse Event Report, Insect Sting, "CIRC# 225, September 24, 1998. "The insect appeared to be a wasp. . . . There was a 0.3 cm diameter erythematous plaque in the right posterior aspect of the neck. . . . An ice pack was applied and patient was observed for several minutes. . . . Part of a stinging apparatus was extruded, with subsequent relief of the irritation." Clinical Research Center, Brookhaven National Laboratory, BNL document.

49. The following are some examples of new contractors: Thomas Jefferson Lab's original contractor was SURA, an organization not dissimilar to AUI, and in its re-competition it took a similar name, Jefferson Science Associates, and its structure included several corporate entities joining SURA. Oak Ridge's new contractor followed a similar structure as BSA, with the University of Tennessee and Battelle (UT–Battelle) partnering, as did other science labs whose location was on federal land. (SLAC, LBNL, and Princeton were exceptions as they are not on federal land. Even the weapons labs—LANL, LLNL, and Sandia—changed to a similar structure, with the distinction of allowing a for-profit status for the contractor, with several partnering organizations each having specific expertise. These new structures have brought new tensions in lab management as the partnering entities did not always share the same priorities for the operation and direction of the labs as their corporate interests differed.

50. Quoted in Charlie Zehren, "New Chemistry: Lab Changing Managers, Mission," *Newsday*, March 2, 1998, A20.

51. "Contractor Change at the Department of Energy's Multi-program Laboratories: Three Cases—Final Report to the Office of Science," January 18, 2001.

52. Howard Schneider to R. P. Crease, email, August 31, 2021.

53. Charlie Zehren, "Science in the Balance," *Newsday*, May 31, 1998, 4; Charlie Zehren and Jordan Rau, "What Progress Left Behind," *Newsday*, November 8, 1998, A4; Jack Sirica, "Radiation Safety: The Great Divide," *Newsday*, November 9, 1998, A20 (with a companion article on same page entitled "Seeking a Source for Cancer from Nowhere"); Jack Sirica, "The Fear at Ground Zero," *Newsday*, November 10, 1998, A7; Jordan Rau, "The Long Road Back: Hard Choices, Heavy Costs," *Newsday*, November 11, 1998, A7.

CHAPTER 8

1. Matthew L. Wald, "Peña Resigns as Energy Secretary, Citing Concerns for Family," *New York Times*, April 7, 1998, A22.

2. Jack Otter, "A Brush with Foul Play," *Newsday*, August 24, 1998.

3. William D. Magwood IV, Zoom interview by R. P. Crease, September 1, 2021.

4. Brookhaven Executive Round Table (the group that succeeded the CWG), July 21, 1998, minutes.

5. "Fish Kill in Peconic," BNL Media & Communications, July 1, 1998.

6. Jordan Rau, "Finger Pointed at Brookhaven Lab," *Newsday*, August 19, 1998.

7. "Lab Backtracks on Plutonium," *Newsday*, July 30, 1998, A28. The Cooperative Sampling Report was issued September 11, 1998, and detailed the sampling locations and results of contaminants. Plutonium was not among the contaminants mentioned.

8. Karl Grossman, "Suffolk Closeup: Mr. Richardson's Record," *Southampton Press*, October 8, 1998, 12.

9. Frank Marotta to Bill Richardson, letter, February 6, 1999.

10. "DOE Gives Antinukes Big Role in Deciding Brookhaven Reactor's Fate," *Science and Government Report*, February 1, 1999, p. 1.

11. Robert L. Park, "Brookhaven: Richardson Looks to the Stars for Guidance," *What's New*, February 12, 1999, http://bobpark.physics.umd.edu/WN99/wn021299.html.

12. Nuclear Regulatory Commission, "NRC Report on Brookhaven Reactor," February 25, 1999.

13. Robert L. Park, "Brookhaven: No Significant Safety Issues Found at HFBR," *What's New*, February 26, 1999, http://bobpark.physics.umd.edu/WN99/wn022699.html.

14. Robert L. Park, "HFBR: Secretary Richardson Looks to the Stars for Guidance," *What's New*, November 19, 1999, http://bobpark.physics.umd.edu/WN99/wn111999.html.

15. Frank Bates, phone interview by R. P. Crease, September 9, 2020.

16. Robert Birgeneau, phone interview by R. P. Crease, September 8, 2020.

17. John Rather, "U.S. Study Rekindles Debate over Reactor," *New York Times*, December 5, 1999.

18. Karl Grossman, "Reactor Is Shut Down for Good," *Southampton Press*, November 18, 1999, 8.

19. "She's a Super Model Citizen," *George*, August 2000, 82.

20. Park, "HFBR: Secretary Richardson Looks to the Stars for Guidance."

21. Bill Richardson, phone conversation with R. P. Crease, August 10, 2021.

22. Bond, who was at the Office of Science and Technology Policy (OSTP) during 1999, heard about decision in the late afternoon and called both Marburger and Lynch. Reaching neither, he incorrectly assumed they were meeting to prepare a response; in fact both were out of town and had not heard about the decision.

23. "Science Takes a Hit," editorial, *Newsday*, November 17, 1999, A48; James M. Klurfeld, "Forbes' Know-Nothings Win the Reactor Dispute," *Newsday*, November 18, 1999.

24. Patricia Dehmer to Peter D. Bond, November 19, 1999.

25. J. Jioni Palmer, "Closing Spurs Powerful Reaction," *The Library*, November 22, 1999.

26. Andrew Lawler, "Meltdown on Long Island," *Science*, February 25, 2000, 1388.

27. John Rather, "U.S. Study Rekindles Debate over Reactor," *New York Times*, December 5, 1999.

28. J. Jioni Palmer, "Closing Spurs Powerful Reaction," *The Library*, November 22, 1999.

29. "Monday Memo," vol. 1, no. 18, to BNL Labwide Broadcasts, November 22, 1999, BNL document.

30. For instance, Barbara Walters, *The View*, ABC, August 6, 2001.

31. Christie Brinkley, phone interview by R. P. Crease, August 4, 2021.

32. Irwin Goodman, "DOE Shuts Brookhaven Lab's HFBR in a Triumph of Politics over Science," *Physics Today*, January 2000, 44.

33. Goodman, "DOE Shuts Brookhaven Lab's HFBR."

RETROSPECTIVES

1. Lyle H. Schwartz, "Lessons to Be Learned from Brookhaven," January 7, 1998.

2. John H. Marburger III, "Imaginary Address to Employees of BNL," in Marburger, *Science Policy Up Close*, ed. Robert P. Crease (Cambridge, MA: Harvard University Press, 2015), 77–78.

3. John Shanklin, "Lessons Learned from the HFBR, or, Advice I Would Give Others Entering a Crisis Situation," undated sheet.

4. Tara O'Toole to R. P. Crease, email, November 17, 2021.

5. Tara O'Toole, phone interview by R. P. Crease, September 14, 2020.

6. Martha Krebs, phone interview by R. P. Crease, October 20, 2020.

7. Jan Schlichtmann, phone interview by R. P. Crease, April 14, 2021.

8. Peter Strugatz to R. P. Crease, email, November 14, 2021.

9. Ernest Moniz, phone conversation with R. P. Crease, November 1, 2021.

10. Scott Cullen, phone conversation with R. P. Crease, April 29, 2021.

11. "Contractor Change at the Department of Energy's Multi-program Laboratories: Three Cases—Final Report to the Office of Science," January 18, 2001.

12. William D. Magwood IV, Zoom interview by R. P. Crease, September 1, 2021.

13. Gerard Tanguay, "'Twas the Night before Restart (1999 Demise of the High Flux Beam Reactor [HFBR])," lyric sheet.

AFTERMATH

1. See, for instance, Paul LaRocco and David M. Schwartz, "How the Government Is Removing the Brookhaven Plume Shows Possible Future for Nassau," *Newsday*, July 25, 2020.

Index

Page numbers in italics refer to figures.